Engineering Ethics

Other Titles of Interest

Engineer to Entrepreneur: Success Strategies to Manage Your Career and Start Your Own Firm, Rick De La Guardia, 2016, ISBN 978-0-7844-1441-5

Economics and Finance for Engineers and Planners: Managing Infrastructure and Natural Resources, Neil S. Grigg, Ph.D., P.E., 2010, ISBN 978-0-7844-0974-9

Becoming Leaders: A Practical Handbook for Women in Engineering, Science, and Technology, F. Mary Williams and Carolyn J. Emerson, 2008, ISBN 978-0-7844-0920-6

The Lead Dog Has the Best View: Leading Your Project Team to Success, Gordon Culp, P.E., and Anne Smith, P.E., 2005, ISBN 978-0-7844-0757-8

Professional Communications: A Handbook for Civil Engineers, Heather Silyn-Roberts, Ph.D., 2005, ISBN 978-0-7844-0732-5

Managing and Leading: 52 Lessons Learned for Engineers, Stuart G. Walesh, Ph.D., P.E., 2004, ISBN 978-0-7844-0675-5

Engineering Ethics

Real World Case Studies

Steven K. Starrett, Ph.D., P.E., D.WRE
Amy L. Lara, Ph.D.
Carlos Bertha, Ph.D.

Library of Congress Cataloging-in-Publication Data

Names: Starrett, Steven, 1967- author. | Lara, Amy L., author. | Bertha, Carlos, author.
Title: Engineering ethics : real world case studies / Steven K. Starrett, Ph.D., P.E., D.WRE, Amy L. Lara, Ph.D., Carlos Bertha, Ph.D.
Description: Reston, Virginia : American Society of Civil Engineers, [2017] | Includes bibliographical references and index.
Identifiers: LCCN 2016058811 (print) | LCCN 2017002923 (ebook) | ISBN 9780784414675 (soft cover : alk. paper) | ISBN 9780784480380 (epub) | ISBN 9780784480359 (PDF)
Subjects: LCSH: Engineering ethics–Case studies.
Classification: LCC TA157 .S767 2017 (print) | LCC TA157 (ebook) | DDC 174/.962–dc23
LC record available at https://lccn.loc.gov/2016058811

Published by American Society of Civil Engineers
1801 Alexander Bell Drive
Reston, Virginia 20191-4382
www.asce.org/bookstore | ascelibrary.org

Errata: Errata, if any, can be found at https://doi.org/10.1061/9780784414675.

ISBN 978-0-7844-1467-5 (print)
ISBN 978-0-7844-8035-9 (PDF)
ISBN 978-0-7844-8038-0 (ePUB)
Manufactured in the United States of America.

24 23 22 21 20 19 18 17 1 2 3 4 5

Contents

Preface

Ethical dilemmas come in all forms of situations for engineers. I remember my first dilemma very well. As a student, I was working one summer for a state department of transportation and was assigned to assist an engineer who was a construction inspector with nearly 40 years of experience. One day he walked into the contractor's supply trailer and picked up a fine radial saw and a long extension cord in good condition. He walked up to the contractor and said, "Hey, these look like pieces of junk that you were about to throw away. I might as well take these if you are going to toss them." The contractor replied after a few seconds, in a low voice, "Yeah, go ahead and take them." The inspector grabbed a few more items without asking on his way to his state truck.

I was young and inexperienced, but I was highly suspicious about what had just happened. Obviously, the inspector was extorting tools from the contractor and the contractor hoped he would receive inspection favors in return. And maybe they had talked privately about a tools-for-favors arrangement. The inspector seemed to have much experience at "shopping" in contractors' trailers and was not acting as a faithful agent or in a manner to uphold the reputation of the engineering profession.

This book is focused around engineering codes of ethics (Additional Resources and Codes of Ethics). We encourage you to read an entire code of ethics prior to reading this case studies book. There are numerous engineering codes of ethics other than the listed ASCE, Institute of Electrical and Electronics Engineers (IEEE), and the National Society of Professional Engineers (NSPE) codes that would provide additional insight as well, for example, American Society of Mechanical Engineers (ASME), Institute of Industrial and Systems Engineers (IISE), Association for Computing Machinery (ACM), American Institute of Chemical Engineers (AIChE), Accreditation Board for Engineering and Technology (ABET), Biomedical Engineering Society (BMES), and American Nuclear Society (ANS).

Our book is intended for practitioners, consultants, government engineers, students, engineering educators, and others who work in engineering. All the cases are based on real situations and organized by the canons of the ASCE Code of Ethics. All the names are fictitious; any similarities to an actual person, company, or situation are coincidental. Each chapter is self-contained, so whatever ethical dilemma you face, you can find insight by reviewing the discussion and cases under the relevant canon. The last chapter presents two larger case studies where multiple canons apply and interact. Throughout the book, each case is followed by questions for you to consider.

I am aware from teaching engineering ethics workshops and courses since the 1990s of thousands of different, specific ethical dilemmas that engineers have faced. Many engineers have faced challenges related to every aspect of every canon in the Code of Ethics. Such challenges have been occurring ever since humans formed civilizations. The formal responsibility of ancient-times engineers to build safe structures dates back to at least King Hammurabi of Babylon (c. 1775 BCE) when he declared, according to Wikipedia, that if a home collapses because of poor construction and kills its owner, the builder shall be put to death. In modern times, ASCE debated forming a code of ethics from about 1893 to 1914, when the first code of ethics was officially adopted. The current focus to protect the health, safety, and welfare of the public was added to the Code of Ethics by the Society in 1977.

Engineering is a vital profession for humanity. We provide technical solutions and opportunities to individuals, communities, towns, cities, schools, companies, and governments. The technically challenging situations that current and future generations face are immense. In creating engineering solutions, it is critically important that engineers maintain high ethical standards, and we are, therefore, entrusted by the public to provide safe drinking water, safe transportation solutions, safe structures to work and live in, and protection of the environment, just to name a few specifics. Our engineering ethics skills are just as important to society as our technical skills. Our hope for this book is that in addition to improving your skills in solving ethical dilemmas, your motivation to uphold the engineering Code of Ethics will increase. Every day, society depends on you, your fellow engineers, and the ethical decisions we all make.

We have created a LinkedIn group titled, "Pursuing Engineering Ethics through Real World Case Studies" for readers to be able to communicate with each other regarding the cases, to provide similar or different cases, to discuss relevant topics, and generally to advance engineering ethics understanding.

Steven Starrett

Acknowledgments

Dr. Steve Starrett's most influential mentor on engineering ethics was Dr. Paul Munger (deceased, civil engineering professor at Missouri University of Science and Technology). Dr. Munger was the chair of the Missouri Board of Technical Professions in the 1980s during the time of the Hyatt Regency disaster in Kansas City, Missouri. He also taught a very interesting engineering ethics course that sparked Dr. Starrett's interest in the subject so long ago. Dr. Jimmy Smith (deceased, civil engineering professor at Texas Tech University, Director of the National Institute for Engineering Ethics [NIEE]) was also influential in shaping Dr. Starrett's career related to engineering ethics.

Dr. Amy Lara would like to thank her doctoral advisor, Gary Watson (Philosophy and Law, University of Southern California, formerly Philosophy, University of California-Irvine), for his careful and thoughtful guidance through the thickets of moral theory, and her department head, Bruce Glymour (Philosophy, Kansas State University), for a great deal of support as she added science and engineering ethics to her research interests. She would also like to thank all her students, over many years of teaching, who pressed her to make abstract theory concrete and relevant.

Dr. Carlos Bertha wishes to acknowledge Col. James Cook, head of the Department of Philosophy at the U.S. Air Force Academy, for his unwavering support, as well as all his colleagues in the department for their valuable insights and contributions.

All the authors would like to thank Araitz Urresti for her artistic contributions and Tara Hoke for her contribution to the discussion on legal topics.

Chapter One

Introduction to Ethics

The Bridge at Crossway Creek

Samantha Cordell couldn't believe what her supervisor had done. As an experienced engineer for a county government, she was overseeing a countywide bridge inspection project. One day, an inspector she respected called to say that he thought the BB–14 bridge at Crossway Creek had to be closed immediately. Samantha met the inspector within the hour and discovered that many of the old wooden pilings were rotted through and didn't even reach the ground. The bridge deflection during traffic was frightening. Samantha exercised her engineering judgment and closed the bridge. Her supervisor, Todd Jackson, who is not an engineer, received many complaints about the bridge being closed. The closure caused a 45-minute detour for local residents, which they found unacceptable. Todd went out to look at the bridge with the local residents, but he did not invite Samantha. He listened to the residents' concerns and decided to reopen the bridge. On his way back from visiting the bridge, Todd stopped by Samantha's office and told her to rush the replacement of the BB–14 bridge at Crossway Creek, and said that he was going to leave it open as long as possible. It was just too big of an inconvenience for the local residents. As Todd left her office, Samantha sat in stunned disbelief.

What would you do if you were Samantha? What should Samantha do? Did Todd do something wrong when he reopened the bridge? Can Samantha remedy the situation? If so, how? If not, what are her options? Situations like the one Samantha is in are hard for engineers. Samantha is trying to do her job to protect the public and now she faces a serious ethical dilemma. Fortunately, the engineering profession has developed numerous codes of ethics, and many people have devoted themselves to studying the perplexing questions raised by ethical dilemmas. In this chapter, we will examine several important theories of ethics, and we will illustrate their respective strengths by seeing how they address Samantha's situation.

What Is Ethics?

In the vernacular, the words *ethics* and *morals* are used interchangeably. Strictly speaking, however, there is a difference. Morals are those rules we govern ourselves by, the principles we live by. Ethics is a field in philosophy that examines value judgments of right and wrong action. An *ethicist*, then, is a person who attempts to answer questions such as, "What is it like to live a good life?" and "What does it mean to say that an action is wrong?" One might say, to put it succinctly, that *ethics is the careful, philosophical study of morals*, because ethics tackles questions such as, "What ought we to do when different moral principles conflict?" and "How can we tell that an edict such as 'lying is permissible' is bad?"

This is a book in what is called *applied ethics*. Ethics are applied when we devote our attention to moral dilemmas that surface in a particular field or profession. Specifically, this book is concerned with moral dilemmas, such as the one Samantha faces, that are common in the engineering profession. Other examples include plan-stamping (e.g., inadequate review of engineering design work), pressure to falsify billable hours (e.g., record the hours on project B but actually finish project A), performing design work related to a topic that the engineer lacks competency in (e.g., a traditional transportation engineering firm obtains an environmental project and has no environmental engineers on staff, so one of the transportation engineers is tasked with performing the design work), and bribes (as in, "What will it take for you to look the other way on this?"). In other words, we focus here on those unique moral dilemmas that engineers face.

Although ethics is a field within philosophy, when the applied ethics of a particular profession are at issue both philosophers and practitioners of the profession need to be involved in generating norms that are suited to that profession. If philosophers with no engineering experience tried to step into the engineering profession and dictate moral rules for engineers, the rules would be unhelpful at best. Engineers are the experts in what dilemmas arise for them and what values are important for the success and integrity of their profession. Because engineering is a profession—a *self-regulating occupation*—a common framework for evaluating moral dilemmas that engineers face ought to be found within the engineering profession.

As mentioned previously, we find that common framework in what the various engineering organizations call codes of ethics: a list of principles, values, commitments, and affirmations by which engineers agree to govern themselves. Of course, a code of ethics is not the only thing guiding our conduct as engineers: we also have laws (local and federal, civil, and criminal), company or department policies, international treaties, and regulations. But following all the governing principles that have been put in place is not the same thing as being moral, because laws and regulations are written to account for the vast majority of situations, and it is impossible to exhaust all possible cases or to articulate and codify all moral demands into these laws and regulations.*

Tara Hoke, general counsel for ASCE and the author of "Question of Ethics" column in ASCE *Civil Engineering Magazine* writes on the matter of laws compared with ethical standards:

> Ethics are a set of moral principles shared by a particular community with the aim of guiding behavior. Laws are rules established by a governmental authority for the purpose of providing order. The two areas often overlap, because both ethics and the law aim to serve universal moral principles of justice, equity, and promoting the public good—but ultimately they approach those goals by two entirely different methods. Ethics will tell you to drive only as fast as you can safely operate your vehicle; the law tells you that you can drive 35 mph on this road, and 45 mph on that one. A highly skilled driver might be able to ethically drive at speeds well in excess of legal limits, whereas an extremely poor driver might be ethically questionable even at speeds that comply with the law. More significantly, a professional may encounter situations where a legal obligation of confidentiality conflicts with an ethical obligation to speak out on a potential harm. Engineers must therefore know both the laws and the ethical principles that govern them, use both to guide their professional conduct—and sometimes, make a difficult choice between the two dictates when faced with an irreconcilable conflict.

* This argument depends on a philosophical position called "Natural Law Theory." One example of such a theory comes from Thomas Aquinas, a thirteenth century Catholic theologian. For the tools required to navigate this particular account, see *Summa Theologica of Saint Thomas Aquinas, Latin-English Edition, Prima Secundae,* Part I–II, Questions 91–97. Natural Law Theory, however, need not depend on religious or metaphysical commitments. One might just accept that not *all* moral commitments—whatever their source may be—are codified in written law, and look for guidance from the various ethical theories articulated in this introduction.

The goal of this book is to unpack the ASCE Code of Ethics, which is divided in a series of canons by way of case studies that illustrate each one of these principles. To conduct this evaluation properly, however, we need to be able to dig deeper than the written code itself. To think critically about a moral dilemma at hand, we ought to be able to point out not only the applicable Code of Ethics canon, but also the ethical theory behind it.† That is what the remainder of this introduction is about: a summary of the major relevant moral theories that help us frame the ASCE Code of Ethics when it is used to evaluate—and resolve—the moral dilemmas contained in a case study. For our purposes, we must place emphasis on the word *summary*. We can provide only the most basic sketch of the various moral theories that can inform our case studies. To help with this endeavor, we will make use of a convenient illustration: the ethical triangle.

The Ethical Triangle

Given that philosophers have been debating ethics for thousands of years, it's easy to see why there would be many, many ethical theories from which to choose. We can, however, classify the majority of these ethical theories under three general categories: character-based ethics, principle-based ethics, and consequence-based ethics. As a simple way of remembering these categories, we arrange them in a triangle.

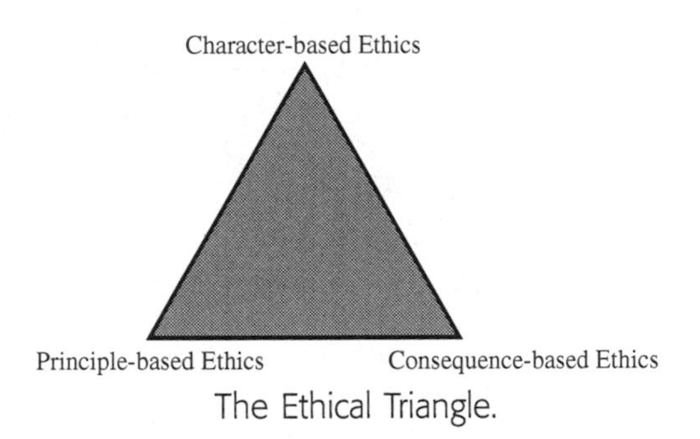

The Ethical Triangle.

Let's look into each category.

† In relation to Thomas Aquinas' version of Natural Law Theory, the ASCE Code of Ethics canon would be an example of Human Law, whereas the ethical theory behind it would constitute our attempt to understand Natural Law.

Character-based Ethics

Character-based ethics, also called *virtue ethics,* was best articulated by the Greek philosopher Aristotle (384–322 BCE), primarily in a work called *Nicomachean Ethics.*[1] As was the case for most of ancient Greek philosophy, Aristotle's primary concern was to answer the question, "What is it like to be a good human being?" Much like it would be reasonable to talk about what characteristics describe a good knife (e.g, sharp, durable, sturdy, comfortable handle) or a good hunting dog (e.g, obedient, good sense of smell, doesn't maul the bird before it takes it back to the hunter), Aristotle thought it is perfectly reasonable to ask what characteristics describes a good—or virtuous—human being. For Aristotle, the *virtue* (in Greek, *aretē*) of a human being came from activities that only human beings could do: reason, deliberate, and study.[2]

To be a virtuous person, however, it is not enough to occasionally reason, deliberate, or study. For Aristotle, a person's identity is derived from what that person does habitually: hence the Greek word for *habit* (*ethos*) gave rise to the word for *character* (*ēthos*). For example, to be able to say that a person is courageous, one would have to point to a pattern of courageous acts, not one fluke instance. Similarly, a person who is said to have a strong moral character is someone who has demonstrated through a life of hard choices that he or she can be depended on to choose the right moral action.[3]

Unlike principle-based and consequence-based ethics, character-based ethics may not be as useful when it comes to discussing individual actions or decisions. Virtue ethics is helpful, however, in framing some case studies if we make a simple adjustment to Aristotle's question about virtue. All we have to do is ask, "What is it to be a virtuous *engineer*?" or "What sorts of characteristics define a good engineer?" Once we answer that question, we can turn our attention to how our answer can help inform a particular case study.

Suppose a young, highly energetic but inexperienced engineer, Craig Arndt, is asked to evaluate the results of a complicated soil toxicity test. We can argue that a key characteristic of a virtuous engineer is to work only on the subjects he or she is technically proficient in. If soil analysis is not yet something Craig has any experience with, we might use virtue ethics to suggest that the right thing to do in this case would be to seek out that expertise and consult with more-experienced (or more-specialized) engineers. In other words, Craig should acknowledge that he hasn't developed a habit of analyzing toxicity in soil samples, so the right thing to do is to find someone who looks at these tests all the time. This action ensures that Craig is doing his best to protect the public: he sticks to what he is good at and defers to others who have a different subject matter expertise.

For our introductory case study, we might say something similar: Samantha is a subject matter expert because she has devoted her career and education to understanding the engineering principles that led her to the decision to close the bridge. She is acting as a good engineer, because she is applying her knowledge to protect the public welfare. She fully recognizes the inconvenience the bridge closure is going to cause local residents. Todd, conversely, does not have that expertise (even though he is her supervisor), so even though his reasons have some merit (he is, after all, trying to remedy an inconvenient situation for the neighborhood), he is wrong to reopen the bridge without consulting with Samantha. Indeed, Todd should defer to Samantha's better judgment and expertise and not open the bridge without her approval.

Principle-based Ethics

Immanuel Kant (1724–1804) is considered the primary proponent of principle-based ethics, also called duty-based ethics or deontological ethics, derived from the Greek word *deont,* meaning that which is binding (i.e., duty). In other words, principle-based ethics deals with what obligations we have, if any.

It is important to understand the context and the era in which Kant was writing, the Enlightenment. During those tumultuous years, putative sources of political and religious authority were being questioned. Most of moral philosophy at the time was structured in what Kant called *heteronomy* (from the Greek *héteros*, meaning other or different), which meant that to understand what we ought to do, we first had to define what we were pursuing (or what our ends were). But because our ends were defined by external sources—generally political or religious authorities—these principles of action would depend on who or what this source was. This situation was, Kant argued, untenable. For us to be able to call a principle truly moral, it must be universal, not source-dependent.

Kant, therefore, flipped the system and encouraged us to *start* with a principle. He then argued that as long as we started with the right principle, we would be doing the right thing *by definition*. In other words, Kant defined good as acting in accordance with the right principle. He called this way of determining a good action *autonomy*, because the decision about what to consider as a principle of action was arrived at autonomously, that is to say, self-selected. Kant thought autonomy was a superior method because this way we are not depending on an arbitrary political or religious authority to define our morality: we are deciding these matters ourselves as moral and rational agents.[4]

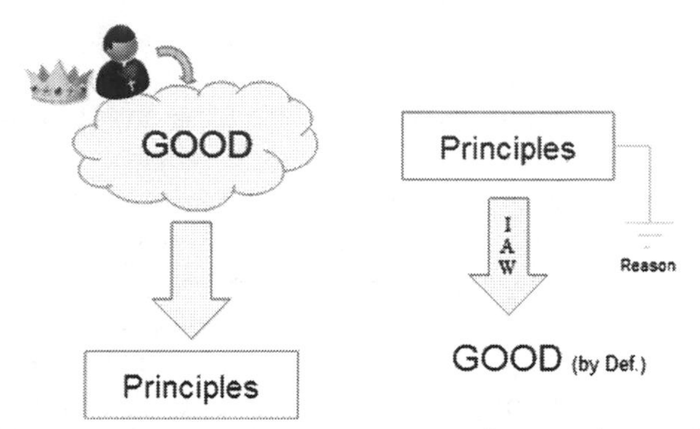

Heteronomy versus Autonomy. IAW = in accordance with.

To the astute observer, this raises the question, "How do we know we are following the right principle?" Kant had an answer: you know that you are following the right principle if it conformed to what he called the categorical imperative: act only on those principles that you would wish to become universal.[5] In other words, applying the categorical imperative requires us to perform a thought experiment of sorts, which we call *universalization*. First, we are to ask ourselves, "What is our dilemma?" We next identify the principle we are about to apply (the personal principle called a *maxim*), and then we are called on to imagine what it would be like if everyone in the world were compelled to adopt the same maxim as a principle of action. If we can approve of such a world, then we have chosen a proper principle and should act accordingly. However, if we could not live in a world where everyone followed that maxim, then we should choose a different maxim.

To act on principles that an individual can will to become universal laws for everyone is to act in ways that can affirm those laws from an objective, impartial standpoint. According to Kant, this way of deciding how to act is equivalent to committing oneself to acting in ways that other rational agents can consent to. If a person is truly impartial, he or she doesn't take advantage of others or manipulate them for self-serving purposes. Instead, one should step back from individual desires and think about what would be acceptable from a more-universal standpoint.

Thus, Kant believed that his categorical imperative could also be stated this way: we should always treat rational beings as valuable in themselves; we shouldn't use them solely for our own purposes. Another way to put this is that we should treat people in ways they themselves could rationally consent to. Sometimes it's easier to use the categorical imperative in this second formulation (called the *Formula of Humanity*), so we will rely on it in several of the case studies in this book.[6]

To understand how to apply this sort of principle-based ethics, suppose a certain engineer is contemplating lying on a report. The lie itself is minor, a fib, that will not alter the conclusions of the report, but it will make him look more competent. Universalization requires that he evaluate the situation systematically. First, he sees that the dilemma he is contemplating is whether to lie on a report. Suppose further that he is thinking about lying: he then says that the maxim that he is about to apply is that it is acceptable to lie to make oneself look good. What would it be like to live in a world where all people governed themselves by such a maxim? One could immediately see that this is an untenable principle to abide by because soon a person would not be able to tell truth from falsity. Conversely, being honest clearly presents itself as a principle that this engineer would want everyone to adopt.

Jack getting a cup of coffee from the breakroom and does not contribute the quarter he is obligated to.

One more example may help put principle-based ethics in context. Suppose another engineer, Evelyn Meade, needs to refill her coffee. She heads to the breakroom, where there is an honor system for coffee, a quarter per cup. She pours herself a cup, puts a quarter next to the machine, and just as she is ready to head back to work, she sees Jack Cranston, a coworker, fill up his cup. And then she sees him walk away. No quarter. She calls him out. "Hey, Jack, it's a quarter per cup, you know?" Jack looks at her funny and responds, "Who cares? Are you going to turn me in? It's only a quarter!"

We can all identify with these sorts of situations. If your reaction is to respond, "Well, Jack, *of course* it's only a quarter. No one is going to miss it, and it won't break anyone's bank. But it's the principle that matters!" You are essentially taking a deontological position. The engineer is saying that rather than the consequences settling whether the action is right or wrong; it's the principle that matters.

Principle-based ethics can also help us sort through Samantha's situation, described at the beginning of this chapter. Todd is acting on a problematic maxim: he is opening an unsafe bridge that could seriously harm the public for the sake of convenience. Can we universalize such a principle, that convenience takes priority over safety? It seems irrational from an objective standpoint: the point of convenience is to improve the quality of life, but doing things that damage or even destroy life itself to improve the quality of life is a contradiction. In this case, it appears as though the residents are consenting to such a trade-off, but they are not actually giving an informed consent. They aren't engineers and don't know the danger the way Samantha does. If they did understand, then as rational beings, they wouldn't want the bridge to be open. Thus, Todd is acting without their true consent, and if Samantha is to follow the Formula of Humanity and truly value the residents as rational beings, she is obligated to protect them from a danger about which they are not aware.

Now we are ready to compare this sort of deontological ethical theory with consequence-based approaches.

Consequence-based Ethics

As the name implies, consequence-based theories focus on the consequences—or results—of our actions. John Stuart Mill (1806–1873) is the primary philosopher we associate with utilitarianism, which is a type of consequentialist theory that calls for us to do that which produces the greatest good for the greatest number. In fact, Mill believed that actions are right in the proportion that they produce happiness and wrong in the proportion that they produce pain. He called this the "greatest happiness principle."[7]

Imagine that Susan Jacob attends a meeting with her fellow engineers and brings a dozen doughnuts. Suppose she gives all of them to her supervisor, Janice Hall. Janice might be pleased with the doughnuts, but wouldn't the overall happiness be increased if Susan gave one doughnut to each participant in the meeting? The greatest happiness principle, then, would say that giving one doughnut to each person is a more-right action than giving them all to Janice. Why? Because it produced more happiness overall.

This approach has a number of interesting implications. For one, consequence-based theories operate under the assumption that there is no such thing as an inherently right or wrong action. Actions are right to the degree that they produce a particular result. Different consequentialist theories, by the way, will propose different things as the result that our actions should produce. For Mill, our actions should produce the greatest amount of happiness as possible for as many people as possible. A variant theory, called *hedonism*, states that our actions should produce maximum physical pleasure. A person who believes that the best result would be the most good to himself or herself would be called an *egoist*.

As you may recall, this is precisely the problem that Kant had with heteronomous systems of ethics: who determines what the best results should be? Mill thought that we should be able to answer this question ourselves. No one would object to certain categories of things that are worth pursuing as ends. In an engineering ethics context, then, we can adapt this approach by taking a closer look at those best results. Much like physicians place a premium on the health of their patients, engineers are responsible for the safety, health, and welfare of the public at large, the sustainability of the environment, and other goals put forth in the Code of Ethics.

In our opening case study, if Samantha were to apply consequence-based ethics to her reasoning, she would say that Todd was wrong to reopen the bridge *because* it put the public at risk and that she should address the matter with him (or perhaps his supervisor) immediately *because* it is her responsibility to protect the public from unnecessary risk.

Consequentialist theories—all of them—are vulnerable to the following concern: they depend on our ability to *forecast* the results of our actions. When we make a decision on the basis of the outcome we want, we are essentially counting on our actions to yield that outcome.

But what if we miscalculate? What if we fail to take all the variables into consideration? What if our actions do result in the consequences we want but also produce additional, negative consequences that we did not anticipate? Therefore, whenever we entertain consequentialist arguments, we should be mindful of just how accurately we might be envisioning the actual results of our actions.

The Moral Spectrum

Facing a choice between right and wrong is not what we call a *dilemma* (we should just choose the right action and be done with it). Dilemmas happen when we must choose between wrong and wrong. Samantha must choose between keeping the bridge open, which allows convenience for local residents but could potentially put them in a great deal of danger, or shutting down the bridge, which protects the public but will result in long delays until repairs can be made, not to mention angering her supervisor (and maybe Todd ends up firing Samantha).

How do we choose between these moral theories? Is it possible for us to be in a situation where the principle at hand is in conflict with the consequences of an action? Are there cases where we can—and should—act against an important principle to achieve an even-more-important consequence? In other words, given the theories we just introduced, who is right, Aristotle, Kant, or Mill?

There are three ways to answer this question, and it may be helpful to think of these answers as being on a spectrum.

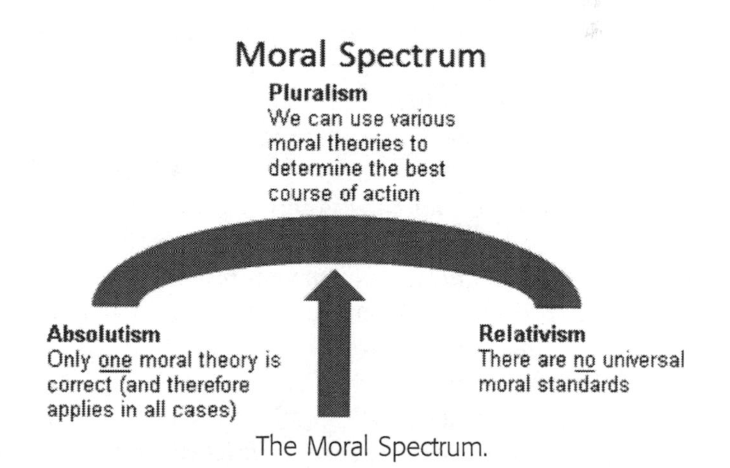

The Moral Spectrum.

At one end of the spectrum, an *absolutist* would say that only one moral theory is correct and that such a theory ought to apply in all cases. One absolutist, Abbot Zirkowski, might say, "The only thing that matters is the principle at hand. The ends never justify the means." When Abbot is confronted with a moral dilemma, he always takes a principle-based approach. Abbot believes that a principle that makes lying permissible is a bad principle, so he would never lie, even if his life depended on it.

At the other end of the spectrum, we have what is called a *relativist*. Moral relativism is a position that argues that there are no universal moral principles, so there is no sense in saying that any of the moral theories depicted on the triangle are right. Another version of this argument, called cultural relativism, takes the position that while some moral standards may exist, they exist only within a particular culture. In other words, suppose that Reilley LaClare, a relativist, lives in the United States and knows that it is not commonly acceptable to bribe a police officer. Reilley, while traveling in Nigeria, notices that the practice is quite common and unproblematic there. She would say that we in the United States would not be able to judge the moral standards of Nigeria (and vice versa), because such standards reside only within a culture. "When in Rome, do as the Romans do," the old adage states.

We should be able to see that either end of the spectrum presents us with some problematic scenarios. On the one hand, an absolutist would be committed to choosing one—and only one—moral theory to guide all moral choices. Yes, principles such as not lying are important, but are we willing to say that there are absolutely no circumstances under which telling a lie is the right thing to do? And, on the other hand, is it really true that there are absolutely no moral standards that cross cultural boundaries? Are we willing to excuse heinous practices around the world under the guise of "when in Rome, do as the Romans do"?[8] There has to be an alternative to absolutism and relativism: the position that all three moral theories explored previously are right—in their own context. And none is perfectly and absolutely right 100% of the time. This middle ground is called *moral pluralism*, which is a position that accepts that it is possible for moral questions to have more than one right answer but that accepting that there may be more than one right answer does not entail that there are no wrong answers.[9]

Suppose we went to a large mall in Dallas and asked 5,000 shoppers, "What is the best movie of all time?" Probably we would not receive 5,000 different answers. Movies such as *The Godfather, Gone with the*

Wind, Saving Private Ryan, and *Schindler's List* may receive a few more votes than movies such as *Star Wars, Avatar, Spiderman,* and *Four Weddings and a Funeral*. There's a pretty good chance that some movies would not be mentioned at all (e.g., *Gigli,* which actually appears on the IMDb list of the worst movies ever made),[10] even if we were to ask 50,000 people. The point is, there are numerous reasonable, defensible answers to our question, "What is the best movie ever made?" And yet the fact that multiple reasonable answers exist does not preclude other answers as being false.

Samantha has a number of options, too. She can challenge Todd directly, appeal to Todd's supervisor instead (jumping the chain of supervision altogether), go to the press, or take some other action. Those options would have a variety of costs and benefits. But the option of doing nothing is not a good one, because it places the public in unacceptable danger. Samantha should rule it out: she must act in some way to protect the safety, health, and welfare of the public.

When we make moral decisions, we might be faced with a number of possible right answers. Each of those answers ought to be backed by some sort of argument. Sometimes the argument might be grounded in consequence-based theories; other times our argument might be based on a character-based theory; and there will be situations that might call for a principle-based approach. The fact that any of the three approaches may apply to any given situation doesn't entail that one theory is always the right one to take, nor does it suggest that any argument presented is just as valid as any other.

A Closing Note

Before closing this chapter about moral theories, a brief comment about our approach is in order. Using case studies to analyze moral dilemmas is called *casuistry*. We employ case studies that are representative of what engineers typically face and use them to show how these various ethical principles—as embedded in the ASCE Code of Ethics—can be applied to help us resolve moral dilemmas. In this context, our case studies are designed to have a resolution rather than a solution. Yes, we want to be bold and call some actions right and other actions wrong, but we should also be sensitive to the fact that some actions can be right at times, or not necessarily wrong. Context and circumstances matter, yes, but we are not relativists! When we believe an action is clearly wrong (misrepresenting a professional license on a resume, for example), we say so. When the only right action to take is to close the bridge at Crossway Creek, we say that, too. As you will see, however, some choices are more difficult to make than others. In those cases, we hope to present a fair analysis of the various options available and the costs and benefits of each. An obvious right or wrong may not be possible, which we think is consistent with the moral choices that engineers face from time to time.

Our goal for this book is to provide you with the tools to help you better resolve the ethical dilemmas that are in your future as an engineer. If you thoroughly consider what you would do in the presented cases, think of similar situation you might have been in, and analyze our solutions to the case study situations, we are confident your ethical problem-solving abilities will be significantly strengthened.

---

[1] Aristotle. (1999). *Nicomachean ethics,* Trans, Terence Irwin, 2nd Ed, Hackett Publishing, Indianapolis, IN. We will use the marginal pagination (sometimes referred to as Stephanus Numbers) to reference the location of the actual citation. This number is the same, regardless of what translation or publisher is used.

[2] *Ibid.* Book X.

[3] *Ibid.* Book II (1103b05).

[4] Kant, Immanuel. (1964). *The Groundwork for the metaphysics of morals.* 3rd Ed. Trans. H. J. Paton. Harper Torchbooks, New York. Here, too, we will be using marginal pagination, which corresponds to the original German edition. Kant discusses heteronomy and autonomy in Chapter 2, 440–442.

[5] *Ibid.* p. 421.

[6] *Ibid.* This formulation of the categorical imperative is sometimes also referred to as the practical imperative and is found on page 429.

[7] Mill, John Stuart. (1993). *On liberty and utilitarianism.* Bantam Books, 144. Unfortunately, Mill's work does not have a convenient marginal pagination.

[8] Rachels, James. (1986), *The elements of moral philosophy.* Random House, New York. This book gives a succinct explanation of the problems with cultural relativism. See especially Chapter 2, "The Challenge of Cultural Relativism."

[9] Adapted from Lengbeyer, Lawrence, "Ethical pluralism: An alternative to objectivism and relativism," Department of Leadership, Ethics and Law, United States Naval Academy, Annapolis, MD, 2003. This article is published in *Moral dimensions of the military profession,* Custom Books, 5th Ed., 41–43.

[10] The list of the worst movies ever made can be found at http://www.imdb.com/chart/bottom?ref_=tt_awd.

Chapter Two

Hold Safety Paramount

Canon 1. *Engineers shall hold paramount the safety, health, and welfare of the public and shall strive to comply with the principles of sustainable development in the performance of their professional duties.*

Something similar to Canon 1 currently appears in the official code of ethics of most engineering societies, and always in a prominent position.[1] It is clearly a central ethical commitment for engineers. The rationale for this canon is twofold. First, the work that engineers do almost inevitably has great impact on the material well-being of people throughout society, including many who do not know what the engineers are working on and who have not explicitly consented to any associated risks. A private individual's choice to engage in risky behavior seems to be a matter of individual freedom, as long as the individual is competent to judge the risks. When our actions impose risks on others, however, ethical principles require that we consider what those people have consented to or could reasonably consent to.

Of course, it is not always possible to obtain explicit consent from others for our risky actions; every time we decide to drive a car, we impose risk on others, yet we cannot ask them all to sign waivers so that we may drive to the grocery store. Thus, we often use proxies for explicit consent: the tacit consent of citizens to the laws in a democratic society, for example, or the assumed consent of an unconscious person to receive medical aid. Because the work of engineers can impose risks on large numbers of people from whom it would be impossible to get explicit consent, engineers must use proxies for that consent. Canon 1 serves as such a proxy. (Operating within the law, although not itself one of the canons, also serves as such a proxy.) We can assume that the vast majority of people value their own safety, health, and welfare and would not consent to reckless disregard for, or even giving short shrift to, their well-being. On the contrary, people would choose that their own well-being be given very high value in other people's deliberations. By explicitly valuing others' welfare under Canon 1, even while knowingly imposing risks, engineers meet the Kantian ethical requirement to act in ways others could consent to. One strength of this rationale is that it appeals to the universal values of respect for others and concern for their well-being. Thus, this canon can serve as a compass point for engineers working anywhere in the world. The universality of this canon will be important in our discussion of the other canons, which sometimes give rise to confusion about what to do when laws and mores vary from country to country.

The second rationale for Canon 1 is more specific to industrialized economies: members of professions bear special obligations to the community in which their profession operates and to the profession itself. There is no universally agreed-upon definition of a *profession*, but many who write on professional ethics define professions as occupations that require "extensive intellectual training" and that provide an "important service in society."[2] If the profession in question requires great skill and has a large impact on public welfare, professional associations likely exist to provide objective standards for the training and certification of practitioners. The more likely it is that unqualified practitioners can do a great deal of harm, the more need there is for formal licensure. For example, although creating beautiful paintings requires a great deal of skill, there's no need for a professional license for artists. By contrast, in many locations people such as physicians or civil engineers must be licensed to practice.

One way general members of the public can be assured that credentialed professionals will not do great harm is for the profession to build concern for public welfare into its professional standards. Once concern for the public is part of a profession's code of ethics, each professional is obligated to the profession and to the public to uphold those standards in return for the professional standing the profession gives the practitioner. Similarly, the profession itself makes a promise to the public to protect its interests in return for the public's support of and respect for the profession. The profession acquires prestige and public trust by operating in an ethical manner.

It is important for members of a profession to understand that they are not merely employees of a business. As professionals, they are obligated to uphold the standards of their profession, even when that requires displeasing or upsetting their employer. They also bear responsibility for the professional decisions they make and cannot claim they were only following orders.

An example of an action that may upset an engineer's employer is called whistleblowing. Blowing the whistle, generally speaking, means making a public accusation concerning misconduct by one's organization. To make the definition more precise, Martin and Schinzinger list some features that characterize whistleblowing:[3]

- Information is conveyed outside approved organizational channels or in situations where the person conveying it is usually under pressure from supervisors or others not to do so.
- The information being revealed is new or not fully known to the person or group to whom it is being sent.
- The information concerns what the whistleblower believes is a significant moral problem concerning the organization.
- The information is conveyed intentionally with the aim of drawing attention to the problem.

In 1974, Turkish Air DC-10 crashed, killing 346 people. The design flaw that led to the crash—a faulty latch mechanism in a cargo door—had been identified by an engineer two years before, but his supervisor told him to drop the matter. If the engineer had blown the whistle, perhaps his actions would have saved the lives of those 346 passengers and crewmembers.[4]

Although blowing the whistle in this case seems like the right thing to do, cases are not always so clear-cut. Engineers who decide to blow the whistle often do so at the risk of losing their jobs (and perhaps harming their entire careers); they can sometimes be perceived as just being troublemakers or as rocking the boat. It short, blowing the whistle appears to be an all risk, no benefit proposition to the engineer. No wonder, then, that there has been some debate about the circumstances under which an engineer is justified in blowing the whistle and, more controversially, whether there are cases when the engineer is morally obligated to do so in order to protect the safety, health, and welfare of the public.[5]

Canon 1 also includes a clause regarding sustainable development. This clause was added in 2009 as the idea of sustainability became a popular way of addressing concerns about the environment. Environmental ethicists have long debated whether we are ethically required to treat the natural environment as intrinsically valuable (valuable in itself, apart from its usefulness to humans) or only as instrumentally valuable. However, this debate need not be resolved before we can justify the ethical soundness of the sustainability clause of Canon 1. Whether or not the environment has intrinsic value, ethics requires us to hold that human beings have intrinsic value, and the health of the natural environment is of obvious importance to current and future generations of humans. Thus, the sustainability clause of Canon 1 flows directly from the requirement to value the welfare of the public; we can think of this clause as simply making explicit something already implied by the requirement to place paramount value on public welfare.

A number of difficulties arise when applying Canon 1 to specific situations. One arises from the wording of the canon itself, which states that engineers must hold public safety, health, and welfare "paramount." What does this mean? Literally, it means that public safety, health, and welfare must be ranked higher in importance than any other consideration: personal profit, corporate profit, loyalty to one's employer, legality, and other professional and personal considerations. Yet, it seems unlikely that the writers of the code meant the word *paramount* to be taken as an absolute. Such a literal code would require unlimited amounts of money to be spent on reducing public risk to the smallest possible level, with no consideration for the diminishing returns of such a venture. Surely what the code means is that public safety, health, and welfare must be given great (but not infinite) importance in an engineer's cost–benefit analysis. But exactly how much weight should public safety, health, and welfare be given? Presumably they should be given a reasonable amount of weight, but what is reasonable? This would seem to be a good place for virtue ethics to step in and provide guidance on how to be an ethical and reasonable deliberator. The lack of such guidance makes Canon 1 difficult to apply.

A further problem arises in valuing public safety, health, and welfare: What about the conflicts that may arise between safety, health, and welfare, or even between different components of each of these complex values? For example, a new and well-designed skate park may bring a community many valuable things: exercise, recreation opportunities for teens, reduced teen crime, and community development. Yet it also brings increased risk of injuries, even fatalities, and the money spent on it could have been spent on other beneficial things. How does one weigh these costs and benefits against each other? If the project engineer is deciding between a more expensive, dangerous, and exciting design and a less expensive, safer, less appealing design, which is better? Which is more ethical?

Because reasonable people can disagree about how to prioritize different values, a democratic society aims to give everyone who will be affected by public projects some voice in how those values should be ranked. This practice reduces the need for the individual engineer to make such difficult judgment calls. In the skate park example, the public has ideally already had a voice in the decision to create a skate park (as opposed to other projects), the decision about how much to spend on the project, and the broad outlines and purposes of the project. The engineer can use these as guidelines in making more specific design decisions.

The questions of how much value to place on public welfare, and how to weigh different elements of welfare against each other, involve value judgments. Most people accept that there can be reasonable disagreement about values, though there are of course limits on what counts as reasonable disagreement. For example, someone who actively wants to harm people would not have a place at the deliberation table. What we sometimes overlook, however, is that there can also be reasonable disagreement about empirical issues. For example, two equally qualified engineers might disagree about the likelihood of a project's causing harm to the public. We never have full information, and our theories are not comprehensive. Because Canon 1 requires us to assign great weight to public welfare, we must predict the consequences of our actions on public welfare; because our predictions carry a certain amount of uncertainty, disagreements will arise and we will not be able to apply Canon 1 unless we know how to resolve such disagreements in an ethical manner. Thus disagreements rising from uncertainty are another source of difficulty in applying Canon 1.

Finally, it is difficult to apply Canon 1 without a clear definition of "the public" whose welfare is to be considered. It is easy to think of the public as simply the people outside the company or agency engaged in the project, but still within the immediate region or country. Using such a definition for the public would be a mistake. First, people *within* the company, from engineers to custodial staff, are also members of the public who must be protected. Second, from an ethical point of view, all people have equal value, regardless of their proximity to the ethical deliberator, so one cannot legitimately ignore the interests of people in other regions or countries. However, it can be difficult to figure out how best to protect the interests of people in different countries. As the following discussion of Case Study 2 shows, the public may set up regulatory practices to protect public welfare, but these practices can differ greatly from one jurisdiction to another. When regulations conflict, or when an engineer believes a regulation is counterproductive, it can be difficult to figure out how to act in accordance with Canon 1. Future generations may also count as members of the public whose welfare needs to be taken into account. The question of how much value to assign to future generations, especially if their interests conflict with the interests of the current generation, is a vexed one within environmental ethics. Without a clear answer to that question, Canon 1 is very difficult to apply to concrete cases where nonrenewable resources are being used or where environmental damage may occur.

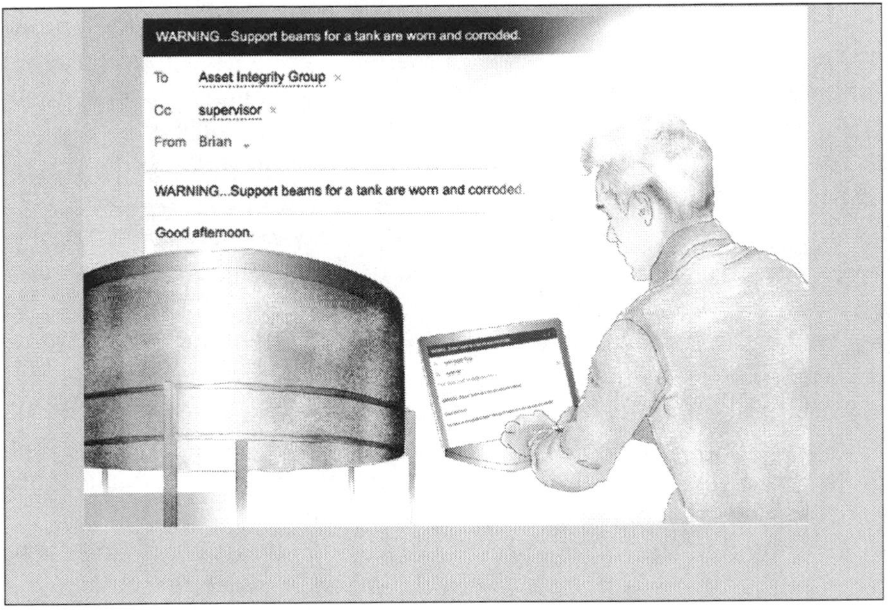

An engineering inspector determines safety hazard at facility.

Case 1. Brian Learns of an Imminently Dangerous Situation

Brian Marshall, a civil engineer early in his career, is providing operations support for a large oil and gas production company. He performs maintenance at one of the company's operational sites and generally works independently. His immediate supervisor is not frequently on the site.

During a routine tank cleaning at the site, Brian learns that the support beams for a tank are worn and corroded and contain several holes. In his judgment, the condition of the beams is a serious issue and should be repaired immediately. However, this tank provides important backup support, so during the time the tank is out of service, there is an increased risk of an unplanned shutdown of the entire plant.

Brian notifies his supervisor by e-mail. After several days, he finally receives a phone call from his supervisor, who tells him the beams are not a serious issue and the tank should remain in service. Brian disagrees with this judgment, and he is also worried about his personal liability because his supervisor's decision has been transmitted by phone rather than in writing.

He decides to call his company's internal Asset Integrity Group, which initially recommends that the tank stay out of service while repairs are made. He also contacts the company's Health, Environmental, and Safety Group, which makes no recommendation on how to proceed.

The supervisor calls Brian again to stress that immediate repairs are not necessary. When Brian talks again to people in the Asset Integrity Group, they change their recommendation and tell him that it is acceptable to postpone repairs for another year. He sends an e-mail to all parties confirming their recommendations and agreeing to postpone repairs for a year. His e-mail gives him a written record of how the final decision was made and who is responsible for it. The written record would greatly validate his effort to provide safety, but any legal action is dependent on how good the attorneys are. So, he may still go to court but win the case in the end (which is not really winning). The tank will remain in service for another year.

Discussion

The central ethical issue raised by this case is the difficulty of making a morally responsible decision under conditions of uncertainty. Subsidiary issues include how to relate in a professional way to coworkers with whom one may disagree, and how to document decisions so that transparency is maintained.

This case involves two of the sources of uncertainty discussed previously. The most obvious is the empirical uncertainty of predicting the effects of one's actions on public welfare. How likely is it that a dangerous equipment failure will occur? If different people or groups reach different estimates of risk, which judgments are more likely to be true? Also, one must not forget to factor in the risks associated with taking action. If repairs end up causing an entire plant shutdown, what effects will that have on the public? This empirical uncertainty leads to disagreements about the facts.

The second source of uncertainty lies in our value judgments. How much value should be placed on public safety versus profitability of the company? This question is complicated by the fact that profitability of the company is also part of the public welfare, because the public has an interest in a healthy economy. Uncertainty in our normative judgments leads to disagreements about values.

In the case at hand, getting clearer about the source of uncertainty would be helpful. Here is where open lines of communication are important. Brian believes that a support beam failure is possible, and that the risk of that failure is significant enough to warrant taking a risk of a plant shutdown while repairs are being made. Others in the company have different views. It would be helpful to know what their views are based on, because this will tell us whether the disagreement is about facts or values. Does Brian's supervisor, Tamir Roache, have some reason to think that support beam failure is very unlikely? Or that such failure, if it happens, is unlikely to be dangerous? This disagreement is empirical and might be resolvable. Tamir has more engineering experience, and he may be making more accurate predictions; explaining his calculations to Brian would help eliminate the disagreement and contribute to the education of the young engineer. Or Brian might have useful information that the older engineer overlooked; discovering this would also be beneficial.

Conversely, the disagreement may not be about the probability of support beam failure, but about the relative importance of other considerations. Perhaps Tamir and the Asset Integrity Group have a broader perspective and are thinking about the long-term costs (to the company and to the public) of a plant shutdown. They may judge that the risks of support beam failure are worth taking to prevent what they consider to be a worse outcome. In this case, Brian agrees with them about how much risk there is (an empirical issue), but disagrees about whether taking the risk is worthwhile (a normative issue). This disagreement can also be overcome by understanding the reasoning behind Tamir's decision. Brian may simply be unaware of the long-term costs of a plant shutdown. Alternately, it may be that there is corruption within the company, and more senior people in the company have become more concerned with profit than public safety. In this case, it is especially important that the young engineer's voice be heard because Canon 1 is being violated by the company.

Because open communication is important for achieving good outcomes under uncertainty, guidelines for ethical communication are an important adjunct to Canon 1. In this case, communication problems existed before the issue with the beams arose and are contributing to Brian's difficulties. He is young and does not have a great deal of experience, and his older supervisor is distantly located and does not visit the site frequently. Communication is usually by phone, which leaves no written record, and communicating via phone instead of in person may make it difficult for Brian to communicate his concerns fully. Tamir should have ensured that there was a robust system of communication in place from the start, but in the absence of that, Brian needs to assert himself and request better communication. He can follow up phone calls with a detailed e-mail restating his position and his understanding of his supervisor's position, or he can request site visits when he judges them to be necessary. It would also be advisable for Brian to keep a daily log of his communications so that he has his own record of what was said by whom and when it was said. The important thing for Brian to remember is that, as a professional, he has more independence and more responsibility than a simple employee. He is obligated to make his best engineering judgments, to express those judgments assertively, and to follow the canons of his profession.

Once communication is opened up, how should engineers talk about their disagreements? Here, the Kantian principle of respect for autonomy can provide good general guidance. Respect for others requires giving them full, truthful information and allowing them to make their own decisions rather than trying to manipulate them into making the decision one wants them to make. There are more specific guidelines that can be especially helpful in cases like Brian's. Engineers are technical problem-solvers and designers, and this perspective gives them an advantage in dealing with disagreement: math, physics, science, and engineering principles form the foundation of their work. Seeing an issue as a technical problem to be tackled together helps people set aside their egos and their desire to defend their own position at all costs.

In the case in question, there is legitimate reason for concern about how likely it is that the support beams will fail and how likely it is that such failure will cause injuries. The disagreement need not be personal: each person is making his or her best judgment given a certain set of evidence about how the world is. If they exchange that evidence with each other, it is possible that the disagreement will dissolve completely.

Even if disagreement remains, each person will be able to see a crucial truth: reasonable people can disagree about the facts under conditions of uncertainty. One need not see one's opponent as morally evil or intellectually obtuse. Often it's simply the case that the evidence underdetermines a conclusion, and we each have to do our best in those circumstances.

So, what of the remaining disagreement? It is precisely because even well-intentioned, intelligent, careful people can disagree that we have developed hierarchies. In the case in question, it is Tamir who has to make the final decision, and Brian should defer gracefully. Note that this conclusion applies only to conditions of uncertainty, where reasonable people can disagree. If the young engineer had good evidence of significant risk and good reason to think his superiors were simply refusing to take risk to the public seriously or were placing profit above the safety of the public, then deference would not be the ethical response. Holding public safety, health, and welfare paramount means putting the public's safety above loyalty to one's own company, above one's interest in promotion, and above one's interest in having a smooth relationship with a supervisor.

Finally, decisions that affect the public, as well as the process of arriving at those decisions, need to be documented, especially when there is disagreement among the decision makers. Documentation is not solely an issue of prudence for the parties involved. There are two substantive ethical considerations involved. The first is transparency. The public has a right to know about decisions that impact its health and safety. The public's right to know of course has to be weighed against other considerations, such as the need for companies to keep some information proprietary, but some weight must be given to the public's right to know. The second important ethical consideration is the need to improve our practices so that safety is enhanced. If an engineering project does end up harming people, it is important to be able to identify the causal chain leading to the negative outcome so that we can make improvements where possible. Thus, Brian was right to create a written record of how the decision to postpone tank repairs for another year was made.

Our conclusion, then, is that Brian should not ignore his worries. He should spend some time identifying the source of his disagreement with Tamir, and if he continues to believe that public welfare is not being adequately protected, he should communicate his disagreement to his supervisor and then up the chain of command if necessary.

Questions

1. What are the technical and ethical responsibilities, in general terms, of Brian and Tamir?
2. How do the differences in responsibilities, and experiences, affect which solution should be implemented for this problem?
3. At what point does an increased risk to public safety justify an engineer's demanding immediate repairs?
4. What are the potential consequences of the two opposing solutions?
5. What virtues do you think are critical for Brian to follow?

Case 2. Errors in Required Records Discovered

Tony Rodriguez, an environmental engineer, has worked for several years for a company that operates massive production plants. Manufacturing so many products creates significant waste, some of which is considered hazardous, requiring extensive documentation, reporting, and proof of proper disposal. One year the EPA selected Tony's manufacturing company for an audit that would cover the entire waste management and disposal processes used at the plants. As part of the audit, three years' worth of records were to be studied and reported on. Tony was responsible for reviewing the thousands of records and developing the associated reports. His extensive review uncovered a few errors in the records. Because an important part of waste management is being meticulous with record keeping, he knew the EPA would not appreciate the errors in the reporting processes. Tony met with his supervisors to inform them of the errors. They also recognized that the EPA would be displeased with these errors, and they feared the consequences. Tony's supervisors then asked Tony to just make this problem go away. They didn't want to hear any more about these errors and they didn't think the EPA should hear about them either.

Discussion

This case raises legal as well as moral questions because Tony's supervisors are asking him to submit falsified documents to a federal agency. Besides the prudential reasons we all have to avoid illegal activity, what moral reasons apply to this case? From a moral point of view, how should engineers relate to governmental agencies that are responsible for regulating company practices but that can sometimes seem intrusive, nitpicky, and even unreasonably punitive? Those being regulated may believe they have an incentive to be dishonest because the regulatory agency will come down hard on any errors reported, even if the errors were unintentional and honestly reported as soon as they came to light. If no good deed goes unpunished, then doesn't it become rational to act unethically and even illegally?

As is often the case in moral reasoning, it helps to step back and remind ourselves of the larger guiding principles. What we have here is a case of manufacturing that affects the general public by affecting the environment itself, so Canon 1 is clearly in play. What complicates the situation is that we're dealing here with a proxy for the public interest. That proxy is the EPA, so it would be helpful to think about the general rationale for agencies like the EPA. This thinking helps us determine the moral status of the EPA's rules and gives us guidelines for ethical interaction with the EPA.

The United States historically has placed great importance on private property rights. The burden of proof is on anyone who seeks to limit the property owner's rights over his, her, or its own property. If I want to paint my living room in pink and purple polka dots, for example, nobody may interfere. However, in the United States, there is also public consensus that interference with private property rights is sometimes reasonable. For example, if I want to pour even a small container of mercury down my kitchen sink drain, the public can interfere. The difference between the two cases is that my actions impose very little risk of harm to the public in one case, but great risk of harm in the other. Yes, even in the living room painting case, there is a nonzero risk that someday you might walk by my house, glance in the front window, and be made mildly nauseated by the hideous colors you see painted there. Perhaps you were walking to a job interview, and your nausea distracts you and causes you to perform badly, and you end up not getting the job. But it would not be reasonable to require people to mitigate that minuscule risk; the risk to public happiness if the government imposed a list of acceptable living room paint colors is much higher than the risk to public happiness if the government does not interfere. Regulating mercury disposal, however, is a different matter. Mercury in the ground or the water poses great risks to human health, whereas the ability to dispose of mercury as one pleases is not of great importance to the average private citizen. Thus, regulation of mercury disposal on private property is reasonable.

How should we regulate those things that may reasonably be regulated? Who decides what activities are to be regulated, how much restriction to place on our activities, and how to monitor our activities? We could of course put these things up to direct popular vote on the grounds that individuals are the best guardians of their own welfare. But besides the onerous burden this would create for people to vote on every possible regulation, there is the more troubling issue that the average individual does not and cannot know how best to protect his or her own welfare. Much of the knowledge required is specialized, and even a very intelligent person could not stay current on more than a fraction of the knowledge needed. Thus, we designate agencies that can collect and analyze the relevant information and form reasonable policies. Trustworthiness is vital to the operation of these agencies because none of us has the individual expertise to verify how reasonable all the agencies' decisions are. We must trust them to evaluate the evidence competently and objectively, and to put adequate importance on public welfare when they make their policy decisions.

A note on terminology: In the United States, the word *law* is reserved for official elements of the legal code as well as for case law that is built up through court decisions. Laws are used to create agencies and dictate their scope and enforcement powers. The agencies then create regulations or rules that explain how the agency is going to carry out the law. Thus, there is a distinction between a law and a regulation. However, speaking philosophically, both official laws and regulations have the general "force of law." That is, they are enforceable rules of conduct that are backed up by the coercive power of the government. The authorization of laws and regulations comes from the same place: the will of the people to govern themselves. Thus, violating a regulation has the same ethical and legal import as breaking a law; in both cases, one is illegitimately acting against the consent of the people.

Just as agencies must set up regulations that are consistent with the laws passed by legislatures and that carry out the people's will, they must also set up reasonable monitoring practices. In an ideal world, laws would be followed simply because they exist, but in the real world laws have no force when they cannot be enforced, and part of enforcing a law is regular monitoring of whether people are following it. Speed limits would have no force if nobody ever checked drivers' speed. Regulations regarding the handling of hazardous materials would have no force if nobody ever checked to see if people are in fact handling the materials safely. There are, of course, many ways to set up a monitoring system;

one has to consider the financial costs, loss of productivity, and intrusiveness of any particular system against its benefits. Balancing costs and benefits is one of the most important tasks a regulatory agency must undertake.

Unfortunately, agencies are not perfect. They are made up of individual, imperfect humans, too. Individuals can be biased, self-serving, ego-driven, power-hungry, incompetent, or simply mistaken. One way agencies try to mitigate this problem is by setting up elaborate systems of rules and oversight so that there is less room for individual discretion; this practice is familiar to us all as bureaucracy or red tape. Bureaucracy has definite costs: it reduces agency flexibility, it slows things down, and it makes the regulatory process more impersonal. The benefits are increased predictability, uniformity, fairness, and accountability.

Keeping in mind the purpose of government regulation, as well as the purpose of uniform, rigorous standards, can help us set aside our frustration and think more clearly about how to act ethically toward government agencies. Covering up errors or lying to agencies is never the ethical choice, because what these behaviors amount to is lying to the public itself (as represented by the government). Less obviously, it is also unethical to undermine an agency's work. The agency represents the public's considered judgment about what risks the public is willing to have imposed on it; to prevent the agency from doing its job is to act in a way the public cannot consent to. Note that this does not mean a company is obligated to do the agency's work for it. In Tony's case, if the EPA asks for three years of records when they could have asked for five, Tony is not obligated to give them more than they asked for. The public is entitled to decide how strict it wants to be about monitoring risky activities, and if it chooses to be lax about monitoring sometimes, that is its prerogative.

Lying to a public proxy is unethical, but the public is not always correct about what's in its own best interest. When the public has made an error, we are obligated to go through publicly endorsed channels to correct that judgment. For example, in the event the EPA hires an incompetent inspector or adopts a misguided set of regulations, there are processes in place to deal with these problems. Indeed, Canon 1 may even require an engineer to take action if he or she has evidence that a regulatory agency is failing to do its job of protecting the public, because it requires engineers to act for public welfare and engineers are in a good position to spot problems with regulatory practices in their area of expertise. The official channels of action may be slow, but it is not okay for an engineer to circumvent the will of the people (as expressed in specific regulations).

Applying these considerations to Tony's case, we can see that his employer is asking him to do something highly unethical: namely, to lie to the public about the company's risky behavior (i.e., the company's production of hazardous waste). The public (through a governmental agency) has decided on a set of rules for regulating the production and disposal of hazardous wastes; it is the public's right to do this because hazardous waste imposes risks on many people beyond the boundaries of the private property the corporation currently owns. Indeed, *hazardous* is just another word for *risk-imposing*, so hazardous waste is an ideal example of the type of thing that may legitimately be regulated by the public. The public, as represented by the EPA, has decided that it would like to have very accurate records of how hazardous waste is handled. It has also decided to enforce this requirement through regular audits and penalties for errors in reporting. If Tony's company knowingly submitted false records to the EPA, the company would be lying to the public about behavior the public has a right to regulate.

The fact that lying to a public proxy is unethical does not mean the public's current set of rules is ideal. In Tony's considered judgment, the EPA may be overly concerned about the particular waste his company produces. It is easy to generate unreasonable public fear about things that scientific evidence shows are quite safe, and that fear can translate into burdensome regulations. Or the particular rules for record keeping may be costly and, in Tony's judgment, ineffective. If so, there are legitimate ways for Tony and his company to notify the public and the EPA about the unintended consequences of their regulations. Lobbying, letters to editors, advertising campaigns, letters of complaint, and other forms of protest are legitimate ways to raise awareness of the problems, as long as they involve honest and informative communication. What is not legitimate is to judge that the public has settled on a ridiculous policy and then unilaterally decide not to follow that policy. This would amount to subverting the basic principles of democratic self-government.

Our conclusion, then, is that Tony should not cover up the recording errors he has discovered. In fact, as an engineer who is obligated to promote public welfare, he ought to report his supervisors' unethical request to people higher up in the company, or perhaps to the EPA if his company is unresponsive and he has exhausted the internal chain of command. People who are responsible for hazardous waste disposal, but who believe they may ignore regulations when those regulations are inconvenient, are a danger to the public.

Questions

1. How important is environmentally related record keeping? What are the potential consequences for not reporting the errors?
2. Why does Tony think the EPA will not like errors in the records?
3. What's the big deal, a few errors out of thousands of entries?
4. If breaking a rule doesn't have significant consequences, then is it reasonable for a person not to follow the rule?
5. What virtues do you think are critical for Tony to follow?

Case 3. Provided Funds Are Not Enough to Build Safe Facility

Susie Nakamora, a civil engineer, has worked for a consulting firm for several years, mostly on airport projects in smaller communities. Because their tax base is smaller, the budgets that small communities have for airports are very limited compared to urban areas. Clients frequently ask Susie to cut corners, to only barely meet the requirements, to specify the cheapest products that will work, and other measures to hold down costs. Susie is worried that with the additional costs of operating and maintaining an underfunded airport, the project would actually exceed the initial costs of building a solid, high-quality airport. In other words, the clients' desire to be very frugal will increase costs over a 10-year period and beyond. She is also worried because a community that builds an airport that lacks the necessary funds jeopardizes the safety of the public.

Questions

1. Why is Susie concerned about these kinds of projects?
2. What do you think is the perspective of the financial manager of Susie's firm?
3. Have you faced a situation where economics and safety were in conflict? How?

4. What are the potential consequences for Susie if she forcefully communicates to a community that the funds needed to do their airport project successfully are just not available?
5. What virtues do you think are critical for Susie to follow?

---

[1] It has not always been the case that something similar to this canon has been included in the code of ethics of engineering societies. See Vesilind, P. A., (1995). "Evolution of the American Society of Civil Engineers Code of Ethics." *J. Prof. Issues Eng. Educ. Pract.*, 121, 4–10.

[2] Bayles, M. D. (1981). *Professional ethics*, Wadsworth, Belmont, CA, 7.

[3] Martin, M. W., and Schinzinger, R. (1989). *Ethics in engineering*, 2nd Ed., McGraw-Hill, New York, 214.

[4] Martin, M. W. (1992). "Whistleblowing: Professionalism, personal life, and shared responsibility for safety in engineering." *Bus. Prof. Ethics J.*, 11(2), 21–40.

[5] See, for example, De George, R. T. "Ethical responsibilities of engineers in large organizations: The Pinto case." and subsequent commentary by Hart Mankin, *Bus. Prof. Ethics J.*, 1(1), 1–17.

Chapter Three

Service with Competence

Canon 2. *Engineers shall perform services only in areas of their competence.*

Canon 2 states a deontic principle forbidding engineers from acting as engineers outside their own area of training and expertise. This principle derives from obligations the engineer has toward three parties: the public, the employer, and the engineering profession. First, public welfare and safety require that engineering projects be carried out competently. Second, those who hire engineers to do work for them need to know that the engineer is not misrepresenting his or her skills. Finally, the engineering profession thrives when there is general trust in its practitioners and in the quality of their training. That trust is undermined by incompetent work, especially when such work is misrepresented by the engineer as being within his or her certified skillset.

Because competently designed and executed engineering projects are so vital to public welfare and safety, Canon 2 could be seen as a sub-principle of Canon 1. Taking public welfare and safety seriously would seem to include a commitment by an engineer to perform only services that he or she is competent to perform. Why, then, add Canon 2 to the ASCE Code of Ethics? What Canon 2 adds is a deontic restriction: a command not to act in a certain way even if you believe the consequences of acting that way will be neutral or positive. In other words, even if an engineer is confident that his or her lack of expertise will not result in public harm on a particular job, Canon 2 says the engineer still should not accept that job. This canon removes some of the burden of judgment from the individual engineer; rather than performing a complex calculation of what would best serve public welfare before accepting a job, an engineer can simply judge his or her own competence to do the job and make the decision accordingly. Because there are objective tools available to judge one's own engineering competence, such as standardized licensing exams, the chances of making a mistaken judgment are reduced, thereby reducing risk both to the public and to the engineer's employer.

The rationale for Canon 2 also derives from the needs of the engineering profession. As discussed in the previous chapter, a profession requires extensive training and skill, often objectively certified by professional organizations. Not all skills require professionalization. I have no need to know whether the sweater I'm buying was knitted by a licensed professional knitter; I can look at the item and decide whether it meets my needs. In fact, requiring professional licensure can reduce competition, innovation, and diversity, so the benefits of professionalization have to be weighed against the costs. However, engineering is clearly a field where the benefits of professionalization far outweigh the costs.

Professions are stable and useful only when their members by and large take the demands of the profession seriously. A profession would be very inefficient if it had to constantly police its members; it needs its members to internalize professional standards. Furthermore, the public's perception of the profession will be formed mostly through interactions with individual members of the profession, not through listening to public relations announcements distributed by professional societies. If individual members of the profession have not internalized its professional standards, the profession will not be perceived as trustworthy.

Thus, it is vital that engineers take seriously their own profession's judgment about their qualifications. When an engineer says, "No, I cannot take on this job, because I don't have the relevant expertise," it contributes to the public's perception of the engineering profession as trustworthy. It is now easier to trust the next engineer who says, "Yes, I'm qualified to do this job." Likewise, when an engineer acts incompetently, it is bad for the profession as a whole.

One implication of Canon 2 is that engineers should be concerned about their profession's judgments of competence. They should stay informed about how the profession certifies practitioners, speak up when that process seems insufficiently rigorous, and provide support for the continuing education of engineers. The profession needs to remain robust, rigorous, and trustworthy so that engineers can continue to receive the benefits of being associated with the profession.

Canon 2 can be difficult to apply to concrete cases. The most obvious difficulty is in determining what counts as an area of competence. Should engineers be required to obtain formal licenses before performing any engineering work, or should a license be required only for certain kinds of work? In the United States, different states have different licensing requirements. Some states (e.g., Kansas) allow licensed

Professional Engineers (P.E.s) to determine the areas in which they are competent. For example, if you have a P.E. license, you could design a water treatment process and circuits for an electrical power control facility if you had competency in both areas. Other states (e.g., California) license individuals as professional engineers in a specific discipline. P.E.s licensed in civil engineering, for example, would not be allowed to design circuits of an electrical power control facility without having an electrical engineering P.E. license as well.

There are some fields where P.E. licensure is not necessary. For example, a P.E. license is not required for industrial design applications, because there is an entirely different process to ensure safety. Many mechanical and electrical engineers, for instance, make products, and those products undergo thorough testing before being made available to the public. These safety processes are governed by abundant laws, regulations, and standards and are managed by numerous regulating agencies (e.g., the United States Consumer Product Safety Commission).

Where legal requirements are looser, individual engineers have to exercise more judgment in order to live up to Canon 2. How does an engineer determine his or her own areas of competence? As we all know, people are not necessarily good judges of their own abilities. Some are overly confident, some overly insecure. Virtue ethics directs us to develop good character traits in this area, humility and objectivity in particular. Humility counteracts our natural defensiveness and helps us face up to our own deficits. Objectivity helps us seek credible evidence, both for what we cannot do and for what we can do, which counteracts insecurity.

Another important virtue is the desire for self-improvement. Does Canon 2 require us to perform only tasks we have done before? That would cause stagnation. It would also be in conflict with Canon 7, which requires continual professional development. It makes more sense to interpret Canon 2 as directing engineers to be especially cautious when expanding their horizons: recognize that you're working out of your comfort zone, consult those with more expertise in that area, and get your work thoroughly reviewed by an experienced engineer (as is the norm at most engineering firms today).

A second difficulty in applying Canon 2 arises from the phrase "performing services." What counts as performing an engineering service? Engineers are frequently asked by friends to provide informal help on various projects. Does Canon 2 apply to the informal work an engineer may do for a friend that is on the fringe of his or her abilities? An engineer also may wish to speak as a private citizen and advocate for various policies in his or her community. Should an engineer refrain from speaking about matters that are outside his or her area of expertise? This would seem to unduly limit an engineer's free speech rights, yet it is not uncommon for naïve members of the public to assume that an engineer has far-reaching knowledge of all engineering-related fields, which might give an engineer extra, and unearned, power in public debate. If an engineer is taken to be expressing a professional engineering judgment when he or she intends only to be voicing a personal opinion, it could end up reflecting badly on the engineering profession.

Once again, it helps to examine the underlying rationale for Canon 2. Beyond protection for public welfare, which is already covered by Canon 1, Canon 2 protects the integrity of the engineering profession. Thus, in making a decision on how to apply Canon 2 to difficult cases, a consequentialist approach is helpful. What are the long-term consequences for the profession if an engineer works on a project for a friend or speaks publicly about a project that is outside his or her area of competence? The consequences might seem small, but psychological research on trust shows us that the consequences could be worse than we think.

Trust is the phenomenon of relying on someone when there is something at stake.[1] One makes oneself vulnerable to another's actions with confidence that the other will perform. When there is no real risk associated with nonperformance, we do not speak of trust. Similarly, when there is no confidence in the other person's performance, there is no trust. Sometimes, of course, it is necessary to make oneself vulnerable even though one is not confident in the other's performance.

Consider a person, Mary Day, who perceives that she lives in a bad neighborhood. She has to leave the house sometimes, even if she is worried that her home will be burglarized or that she will be mugged. Does she exhibit trust when she leaves her house? No. The lack of trust makes itself visible in her actions: she locks multiple deadbolts, carries Mace, and watches others suspiciously. Through her defensive actions, she may end up preventing the loss of her property, but surely she would be better off in a neighborhood where she could trust others to respect her property. Her neighbors would be better off if they had Mary's trust. They could talk freely with her and benefit from her trusting participation in the community.

Similarly, the engineering profession can flourish only with public trust. Engineers work on projects that are important to the public, where there is a lot at stake, so one element of trust is in place. But is the confidence in place? The less public confidence there is, the more regulation and oversight will be imposed, which would reduce engineers' autonomy and flexibility, and ultimately lower their morale. Engineers are better off when they have the public's trust and can exercise their own judgment in their engineering decisions.

Unfortunately, the confidence element of trust is difficult to build up and very easy to destroy. This phenomenon is due, once again, to human psychology. We are all prone to negativity bias, which is the tendency for negative events to be more salient to us than positive events are.[2]

Return to Mary's situation. One scary experience of being menaced or having her house vandalized is much more likely to stick in her mind, and make her feel unsafe, than the dozens of neutral or positive experiences she has in her neighborhood every day. We are also prone to group-attribution error, or the tendency to take an individual's behavior and attitudes as representative of the group to which that individual belongs. When Mary is menaced by one person she recognizes from her neighborhood, she may suspect that all of her neighbors would harm her if they could. She develops a generalized feeling of danger toward the whole neighborhood. It is obvious how these two cognitive biases apply to maintaining public trust in the engineering profession. One disaster will be noticed and remembered much more vividly than hundreds of successes. One unethical engineer can make people feel negative toward the whole profession. These risks must be taken into account when an engineer is considering informal work or public speech that is outside his or her own area of competence.

Case 1. Local Firm Pursues All Types of DBE-Required Projects

Mohammed Jirad is an engineer for a small consulting firm, Four County Engineering, LLC. There are about a dozen people at the company who work on a wide range of projects. The firm has a Disadvantaged Business Enterprise (DBE) certification so the company can submit proposals for government projects that require a certain portion of the overall project to be completed by a DBE-certified company. There are not many DBE firms in the area, so Four County Engineering submits proposals for almost any kind of project that has this requirement, and DBE work is a substantial amount of the work that the firm completes. Mohammed's skills are mostly in transportation engineering, although he has worked on a number of other projects. The owner, Kelly Haksar, has just happily informed the staff that Four County Engineering has received a DBE-related project; she assigns Mohammed to be lead engineer. The project is the interior renovation of a building at a local college. Mohammed's first thought is, "I don't know anything about major building renovations." He tells Kelly that he has little experience in building renovations, and she replies, "Oh, I am sure you can handle it."

Discussion

Mohammed is in a situation that is common for engineers working at small firms: his firm can thrive only by taking on a wide variety of projects, but the firm is too small to have a stock of specialized engineers for every project taken on. All engineers in the firm will need to have broad skills to take on all types of challenging new projects. This situation can create pressure to violate Canon 2.

What would Canon 2 tell Mohammed to do in this situation? He needs to look very closely both at the project he is being asked to do and at his own training. Receiving feedback from others would also be useful. His boss has just told him in a somewhat breezy way that she is sure he can handle the job, but it would be useful for Mohammed to hear more from her about why she thinks that. Does she have an objective basis for making that judgment? Why did she pick him as lead engineer? As the owner of the company, she has a broader perspective and may see ways that the renovation project is similar to earlier projects Mohammed has worked on.

Mohammed should also look closely at the project. Often when we put something in a general category, such as building renovation, we fail to notice that the project has many component parts that overlap with the components of other projects. If Mohammed thinks about the components of this project, he may see many connections to previous work he has done. Also, he could identify individual elements of the project that could be subcontracted out to specialists, although the cost may be prohibitive. On the other hand, if Mohammed is so inexperienced with building renovation that he cannot even identify the component parts of this project, it would be a good indicator that he should not be the lead engineer on the project.

After talking to his boss and looking more closely at the project, Mohammed should consult with his engineering colleagues, both within and outside the company. An important part of professional development for engineers is forming and maintaining connections with other engineers so one can get objective feedback on one's work. Networking outside one's firm is especially important for engineers working in small firms. If Mohammed has good connections, he can talk to people with more experience in building renovations and find out what problems he is likely to encounter on the project.

After he gathers his data, Canon 2 directs Mohammed to make an honest, objective judgment about his ability to be lead engineer on this project. To be objective, he will need to take into account and try to correct for his own biases. He likely has a bias toward making his employer happy, because his employer is the source of his income. So Mohammed needs to think about whether he might be engaging in wishful thinking when he looks over the project and thinks, "That doesn't look so hard after all." He may also have biases regarding his abilities, either toward a lack of self-confidence or toward overconfidence. If he tends toward excess humility, he should work to remind himself of his previous successes in learning new things. If he tends toward overconfidence, he should remind himself of times when he discovered that gaining expertise was harder than he had anticipated.

Once Mohammed has made a judgment, he needs to follow through on it. If he judges he is able to be lead engineer, he still needs to delegate those parts of the project he cannot take on by himself, and he needs to consult with others on parts of the project he has less experience with. Conversely, if Mohammed judges that he is not competent to be lead engineer on this project, Canon 2 instructs him to decline the job assignment or seek additional training. This course of action may lead to bad consequences for Mohammed, but obligations to the public, his employer, and his profession override considerations of personal gain.

Questions

1. What is Mohammed's perspective on this project, and why?
2. What is Kelly's perspective on this project, and why?
3. What are the client's expectations of the Four County Engineering firm on this project?
4. How do you define areas of competence of an engineer?
5. Describe four potential scenarios that could occur as a result of Four County Engineering performing this work.
6. What do you recommend Mohammed do, and why?

Case 2. It's a Safety Violation Only If It's Documented

Jennifer Griffin, P.E., an electrical engineer at a large industrial company, has been assigned to fix an electrical problem that recently occurred at one of the company's chemical plants. Upon Jennifer's investigation into the situation, she determines the following:

- A large power conditioner caught fire because installation did not meet local code requirements; it lacked the use of a neutral, the conditioner was undersized given the electrical load demands, and inadequate overcurrent protection was provided. The conditioner overheated and caught on fire. The fire caused the fire suppressant system to turn on, which caused damage to other equipment.
- The electrical installation was designed and installed by Mark Oren, an individual without a recognized engineering background or formal education.

Jennifer redesigned the electrical installation to meet local code requirements and to ensure it would function properly and not be a safety hazard. Upon completion of the repairs, she wrote a report describing the incompetent design of the electrical installation. She also highlighted the fact that Mark had designed and installed the facility and that he was unqualified for such engineering work (i.e., he lacked proper training and experience). Jennifer's supervisor was uninterested in her investigative report and promptly ignored it. The company had decided not to recognize the event as an official safety issue and just wanted to forget it happened. Jennifer was concerned that Mark would continue to perform engineering services for the company and that other improper engineering designs would be constructed.

Discussion

Jennifer is in a difficult situation. She has been asked to solve problems that are within her area of competence, but in doing so she has uncovered evidence that a fellow employee may be operating outside his area of competence. What are her obligations when she suspects someone else is in violation of Canon 2?

First, Canon 2 requires Jennifer to be sure that she is competent to make the judgments she is making. She is a licensed professional engineer, so she has objective evidence that she is competent to redesign this installation and to diagnose the cause of the original problems. However, does she have competence to make a judgment about someone else's electrical engineering competence? In most areas of expertise, we recognize a distinction between the skills and knowledge required to be a practitioner, the skills and knowledge required to be a contractor, and the skills and knowledge required to *teach* the practice and make judgments about others' abilities. Generally, teaching and judging other practitioners requires an additional layer of expertise. We also may require even more expertise before someone is qualified to formally certify another person's competence. A person who is a licensed professional engineer is not thereby allowed to issue professional licenses to others. State licensing boards grant professional engineering licenses when an applicant proves he or she has the required formal education and practical experience, and in most situations has passed two standardized engineering tests and received recommendations from licensed P.E.s.

In Jennifer's situation, she is not being asked to make a formal judgment about Mark's abilities. Rather, she has come across a troubling situation where an installation was clearly designed and executed incorrectly, perhaps even incompetently. It is well within her range of competence to make the judgment that this project was not well designed. However, because Jennifer is not Mark's supervisor, it is not her role to pass judgment on his overall competence. Furthermore, even if she is qualified to judge Mark's overall competence, she may not have enough evidence yet to justify a general judgment. It is possible that Mark simply made some mistakes in this installation, but that he is overall skilled in electrical engineering.

Jennifer's obligation under Canon 2, then, is to issue a judgment that is in line with her competence. She is obligated to notify her company about her worries regarding Mark's performance on this project. She should also express her worries about Mark's lack of formal training, because this is a good indicator of lack of competence. She is not in a position, however, to issue a formal judgment about Mark's overall competence. If the company ignores her worries, Canon 2 does not give her any additional obligation. However, because Mark's incompetence may affect public welfare, Canon 1 applies and tells Jennifer to report the issue to the state engineering licensing board.

Questions

1. What's so important about a person's having a formal educational background before he or she can be an engineer?
2. Why is Jennifer concerned about Mark's continuing to perform engineering work?
3. Is Jennifer's concern about Mark's ability justified?
4. What is the significance of the problem's not being declared a safety issue?
5. What are Jennifer's options if her company doesn't support her concern about Mark's work?
6. Was this problem partly caused by the industry exemption? Should more people performing engineering work be required to be licensed?

The rest of the story is that Mark was soon promoted to engineering manager. Jennifer eventually left this company to pursue other opportunities.

Case 3. Family Friend Asks for Quick Review of Fire Alarm System

Dan Sturrich received his P.E. license a few months before. A close friend, Jose Carrera, approaches him with an idea. Jose owns a small apartment building where he is putting in a new fire alarm system. It is required by the local code that a P.E. design and stamp the construction plans for fire alarms for commercial buildings. Dan is a talented electrical engineer, but he has not previously worked with fire alarm systems. He works in a totally different field of electrical engineering. Jose is very frugal and doesn't want to pay an engineering firm for what he believes is just a rubber stamp on the very detailed plans that were provided by a vendor. Dan's family is good friends with Jose's family; they often socialize together, and Jose is persistent in asking Dan to just look it over and stamp the plans. Dan doesn't care about the few hundreds of dollars Jose is offering. The only reason he is considering this work is because Jose is a family friend.

Questions

1. Why is Dan reluctant to review the plans and stamp them?
2. What are three possible outcomes of this situation?
3. What is the significance of others' (engineers and nonengineers) telling Dan he is qualified to review the fire alarm system plans?
4. What would you do if you were Dan?
5. If you were Dan, how would you feel if you stamped the plans?
6. If you were Dan, how would you feel if you refused to stamp the plans?

---

[1] Baier, A. (1986). "Trust and antitrust." *Ethics,* 96, 232.
[2] Baumeister, R. F., Bratslavsky, E., Finkenauer, C., and Vohs, K. D. (2001). "Bad is stronger than good." *Rev. Gen. Psychol.,* 5, 323–370.

Chapter Four

Issue True Statements

Canon 3. *Engineers shall issue public statements only in an objective and truthful manner.*

If you ask most people what engineers do, they will say that engineers solve problems. They picture the engineer designing solutions, crunching numbers, inspecting blueprints, and completing projects. They don't think of the engineer as primarily a communicator, even though spoken and written communication constitutes a large proportion of an engineer's responsibilities. Engineers are not isolated individuals, puttering around in basements, imagining possible solutions to theoretical problems. They are people who are called on by the community to design and communicate workable solutions to real problems. Their solutions are made to be used and relied on by others, possibly even by thousands or millions of people. Thus, communication is an integral part of an engineer's work, and the ethics of communication is a vital part of engineering ethics.

An engineer's communication takes many forms, from informal verbal statements to individual colleagues, students, and supervisors, to creating engineering drawings, to interpreting construction plans for contracts, to formal testimony in a court. Traditionally, humans have drawn moral and legal distinctions between verbal and written communication and between informal and formal communication. These distinctions greatly complicate the application of Canon 3, so it will be useful to discuss the general ethics of communication before looking at specific cases that arise in engineering.

The injunction to speak truthfully is a central command in all ethical theories, although it is justified in different ways. *Kantianism* focuses on the way lying displays disrespect for others' autonomy by undermining their ability to reason for themselves from true information. *Utilitarianism* focuses on the importance of trust for social cooperation; lying breaks down that vital trust. *Virtue ethics* focuses on the character trait of honesty, the way it displays a person's commitment to being a full participant in the human community rather than a cynical manipulator.

The ethical requirement here would seem to be clear cut. However, ethicists since the time of Plato have recognized that figuring out what counts as honesty versus dishonesty is very difficult in real life. For example, a lie cannot be defined merely as the intentional stating of something false, for acting, writing fiction, telling jokes, and countless other benign activities all involve intentionally stating falsehoods. Even if we narrow the definition down to statements of falsehoods that are meant to manipulate or trick the hearer, we won't have captured dishonesty in the moral sense, for many such "lies" don't strike us as immoral. Planning a surprise party may involve tricking the recipient, but it hardly seems disrespectful or harmful to social trust. Telling children there is a Santa Claus also seems benign to many people. Furthermore, lying to protect national security or to save lives strikes most people as morally unproblematic.

These considerations show that it is difficult to define lie in a way that is narrow enough. We don't want the definition to include actions that we intuitively would not classify as immorally dishonest. There is also a problem in the opposite direction: that our definition may be too narrow and leave out many actions that do seem dishonest. First, one can say something that is literally true, but that one knows will produce a false belief in the hearer. For example, you might tell a colleague, "Yes, I'll be there tomorrow," knowing that you're planning to leave work before lunch and your colleague needs to meet with you in the afternoon. What you told your colleague was true, but you've been dishonest. Second, one can convey false information without actually saying anything. Body language can convey information, as can simply *not* saying something. For example, if a police officer at the scene of a crime says, "If you witnessed any part of this crime, please speak up," and you are a witness but don't speak up, you are, through your inaction, lying. Kant famously said that all lies, even a lie to a murderer at one's door, are forbidden, but he applied this rule only to actual verbal or written statements. When it came to body language or lying by implication, he endorsed a *caveat auditor* policy: if the person hearing my verbal communication interprets my gestures as indicating something that's actually false, that's his problem.[1] This formulation hardly seems an adequate account of lying.

The way that our society has dealt with these complications is to carve out an arena of communication where the rules of honesty are more precise and explicit, and to leave the rest of our communication a matter of personal character. We leave it up to individuals where to draw the line between honest and dishonest communication with their friends and family. We're all familiar with people who habitually tell you what you want to hear because they value kindness over honesty, other people who can be almost brutal in their honesty, and yet other people who are very private and tend to reveal as little as possible. It's not clear that one of these characters is better than the other. However, we cannot tolerate such diversity in all settings. When witnesses are providing testimony, for example, we need to know that they're not shading the truth, trying to say what we want to hear, or omitting crucial information.

Thus, we have rules for communication in courts, communication from doctors to patients, formal communication from licensed professionals to clients or the public, communication from teachers to students and to the public, advertising communication, and many other types of communication where it is vital that people receive accurate information from which they can reason. For example, if a doctor lies to a patient, the patient cannot make good decisions about his or her own care, which could have disastrous consequences. Telling the patient only what he or she wants to hear is not an ethical option.

The rules for these more formal kinds of communication have the ultimate aim of ensuring that people have access to accurate information that they need to make important decisions. Because nonverbal communication can also interfere with receiving accurate information, these rules cover that as well. For example, an advertisement that contains only truthful statements but that also contains very misleading images can be deemed unethical or even illegal by the rules of honest advertising. Omitting crucial pieces of information can prevent good reasoning on the part of the hearer, so the rule to provide the *whole* truth is included in some contexts, such as sworn testimony.

Because engineers communicate information that is of great importance to other people's decisions, there is good reason for the engineering profession to insist that engineers follow these more demanding and precise rules of honest communication in their professional communication. Thus, Canon 3 says that in their professional communication, engineers are to speak in a "truthful manner." This is a broad requirement that could in principle include not only the literal meaning of the sentences engineers utter or write, but also the implications of what they say, for example, the nonverbal communication that accompanies their statement, images and drawings, and information that is *omitted*. Engineers may even be responsible for foreseeable and preventable misinterpretations of what they say, because misinterpretations can interfere with the hearer's ability to make good decisions on the basis of what an engineer has conveyed. For example, purposely describing a situation in the most complex way possible so as to confuse the public or a jury would violate this canon.

Canon 3, however, requires not only truthfulness but objectivity as well. What does it mean to communicate in an objective manner, and why is it important? To be *objective* is to be unbiased, to be influenced as little as possible by what you wish were true or want others to believe is true. Objectivity is crucial in science because we cannot know if a piece of evidence supports a conclusion unless we know that the evidence was gathered accurately and the reasoning from the evidence to the conclusion was performed logically. Biases in our reasoning can interfere with both data collection and reasoning from data. In fact, the scientific method was designed precisely to counteract common biases, such as confirmation bias (looking only for evidence that supports your theory) and assimilation bias (seeing the evidence that supports your theory as being stronger than the countervailing evidence).

The objective manner in which science is conducted can be extended to the communication of that science. Even if an investigator has performed a careful, objective study and has reached a well-supported conclusion, that person is still subject to a great deal of bias in the way he or she discusses the study with others. In the first place, the investigator may not be completely happy with the results of the study. For example, he or she might have a financial stake in the truth of a certain claim, but unfortunately the study shows that the claim is false. Secondly, he or she may have a strong interest in how the findings are made use of by his or her audience. Perhaps the investigator is presenting the results to a governing body, and he or she hopes they will decide to take a particular course of action, but there's a good chance they'll decide to do something else instead.

Such personal interests can interfere with communication. Every time we communicate, we have a huge number of choices to make about what we include, what we leave out, what we emphasize and deemphasize, and what vocabulary we use. Our choices are complicated by time constraints, our own communication abilities, the level of understanding the hearer has, and other factors. Compare the statement, "Under most conditions, this land development will not cause an increase in flooding," with the statement, "This land development will cause approximately a one-tenth of one foot rise in the floodplain during a 100-year storm event." One statement is focused on the most typical river conditions, and the second is focused on a flood situation. One is very general and one is more precise. Both statements may be true, but they may have quite different effects on the people who hear them. It's easy to see how personal interest might influence one's word choice, even if one is committed to being truthful.

Recognizing these biases is where objectivity enters. To communicate in an objective manner is to try not to be influenced by personal bias in one's communication. One's primary goal is to make sure the people one is communicating with get the information they need to make a good decision by their own lights, rather than to get them to reason in a way that will serve the speaker's interests. It requires being very aware of one's own interests so that one can look carefully at how one's communication is being shaped by those interests. The cases below will highlight how difficult it is to implement this ethical requirement.

Canon 3 can still be difficult to apply even if truthfulness and objectivity are well understood. The first difficulty arises in trying to figure out what counts as a *public* statement. There are clear-cut cases, such as providing expert witness testimony in a court of law, stamping a plan with one's official seal, or submitting a report to a government agency. But what about making nonengineering-related statements in public, such as speaking in a town hall meeting about some political issue? Surely Canon 3 would not prohibit the engineer from speaking *subjectively* in such a forum, for that would overly restrict the engineer's life as a private citizen. This is a good place to bring in the idea of wearing different hats. Making clear when you are wearing your engineering hat and when you are wearing your private citizen hat will go a long way toward helping the audience reason accurately about what you are saying. An official expert statement should play a different role in the public's deliberation than the statement of a personal preference. The former carries extra weight and can outweigh the opinions of others who are not experts, but an engineer's personal preference should carry no more weight than anyone else's in public deliberation. Social media complicate the public–private distinction a great deal, and one of our cases below will address this issue.

The other difficulty in applying Canon 3 is that it is not clear what our ethical responsibilities are when we are communicating with people who are not fully competent to reason well from the information we are communicating to them. This issue is important for engineers because they must frequently communicate complicated and technical information. Often the more precise and unambiguous the engineer tries to be in his or her statements, the more audience members he or she will lose. To aid audience understanding, an engineer may turn to similes and metaphors. For example, electrical voltage can be described as being similar to water pressure and electrical amps can be compared to water flow, or molecules of gas might be described as tiny "balls" that "bounce" against each other. By definition, a metaphor is not literally true, and when misused, metaphors can seriously mislead. Does Canon 3 therefore forbid their use?

More troubling, even when an audience understands the technical information being conveyed, the audience members' own cognitive biases may prevent them from reasoning well about that information. In addition to confirmation and assimilation bias, framing effects (reasoning differently depending on how the information is conveyed), anchoring effects (giving more weight to information that is especially vivid or simply presented first), and many other biases can interfere with what the engineer believes is good reasoning. May an engineer purposely make use of these biases to nudge an audience toward better reasoning? This is a difficult question that cannot be fully addressed here, but we recommend following the general principle of objectivity: as far as possible, the engineer should try to convey information in a way that helps the audience members reason for themselves and reason in a way they would consider to be good; the engineer should not try to get the audience to reason in a way that serves the engineer's personal interests.

Case 1. Creating Engineering Solutions That Maximize Damages for Legal Case

Sam Granger graduated from a university with a degree in civil engineering and went to work for a solo engineering firm. Edward Weill, the owner, had a strong background and interest in traffic engineering. In the few years Sam had been working, he had designed several street renovation projects and lane expansion projects, improved traffic routing projects, and worked on other street-related projects.

Times were tough, so the very small consulting firm accepted any projects it could. One day Edward informed Sam that he had gotten an expert witness project that was focused on site development and property value. Neither Edward nor Sam had a background in these areas. Edward was serving as an expert witness for the plaintiff, who was in a condemnation battle with the state government. The state had offered private landowners compensation for an approximately 10-foot strip of property so that the local road could be widened. The landowners refused the offer and the state was proceeding with condemnation hearings. The landowners' position was that the state was not offering a fair price for the land.

Edward assigned Sam to develop potential site plans for the private land that maximize the potential financial loss that would be caused by the loss of the 10-foot strip of land. Edward informed Sam he was being compensated a fixed amount and a percentage of whatever increase in compensation from the state resulted from his "expert witness" testimony. Edward made it clear to Sam that he expected the analysis to show an extremely high loss, way beyond a reasonable value.

Discussion

Sam's dilemma is a complex case to which several canons could apply. Canon 2, regarding working only in one's area of expertise, may come into play. Canon 4, requiring the avoidance of conflicts of interest, also seems relevant. Most saliently, though, Sam is in a situation where he is being asked to perform engineering work for the purposes of communication. Canon 3 is thus very relevant and will be the focus of this discussion.

The first thing to note is that the type of communication to which Sam's work will contribute is formal, public, and well defined. There are explicit rules for any testimony in a court of law, and those rules reflect the public understanding of the purpose of such testimony. The court (which represents the public interest in justice) expects the truth, the whole truth, and nothing but the truth. It even requires those giving testimony to take an oath to that effect.

One complicating factor, however, is that expert witness testimony is always about something debatable (or there would be no need to bring in an expert to testify). Edward will not be testifying on settled facts, but on the value of the land, which is something about which other experts could disagree. Thus, there is room for individual judgment about what exactly the truth is, which leaves room for personal bias to creep in.

A second complicating factor is that Sam is not the person who will be taking an oath and serving as an expert witness. He is to provide a potential plan for developing the land, which Edward will then use in his testimony. Sam might think, then, that he is not lying to anybody by creating a site plan that maximizes potential loss for the landowners. Certainly he is not lying to Edward, who is asking him to do it. Nor is he directly lying to the court. If Edward chooses to present misleading information to the court, it is not Sam's fault.

A third complicating factor is that Sam is being asked to create only a *potential* site plan. The purpose of the plan is to show how the land could conceivably be used. It is not to create a real plan that an individual or company would be willing to pay for or that represents the most sensible, economical use of the land. Sam's task allows for a lot of latitude. He can generate an outrageously expensive plan for the land without technically lying. He can say in all honesty that his plan is in fact *one* way the land could be used, even if it's highly unlikely anyone ever *would* use the land this way.

Canon 3, however, does not require mere technical honesty. It requires objectivity, which is a very demanding standard. Because Sam is being pressured by his boss to produce a certain kind of result, he is inevitably going to be reasoning in a biased way. For both self-interested reasons and for reasons of loyalty to his employer, he wants to produce something that will make Edward happy. This desire will influence everything he does, from collecting data on the land, to imagining possible uses of the land, to choosing how to communicate a plan for the land. In order to be objective, he needs to try to counteract these biases.

A good way to counteract such biases would be to seek third-party advice from someone who is not employed by Edward and who has no stake in the outcome. If Sam can run his plans by such a person, he can get objective feedback on what would count as a reasonable site plan. He can also show that person his final product to see whether he is communicating in an objective way. Somebody who lacks the biases Sam is subject to in this circumstance will be able to see those biases at work more easily than Sam can.

All these steps would be a lot of work to go through, however, and may not even be possible. How would Sam pay this third party to review his work? If he's looking for free advice instead, will the advice be of high quality? Another option would be for Sam to review recent land developments in the area. There should be public access to plans at the local government zoning office, and comparing recent developments with his plan could give Sam a reality check.

Is all this work really necessary? Why should Canon 3 be so stringently followed in a case like this? An alternative way of looking at the situation would be to put the burden of determining what is objective, reasonable, and factual onto the court itself. The United States has an adversarial court system. Each litigant has an equal opportunity to present the best case possible for its side, and the court (judge or jury, depending on the type of case) renders a decision after hearing each side's argument. Each side is biased toward its own case, and the objectivity comes from the process of having a third party hear both sides and make an independent judgment. If Sam is preparing materials for an expert witness who is hired by one side, what's wrong with his preparing those materials with a strong bias toward that side's case? The other side will present counterevidence, and the court can sort out for itself whether the testimony is credible and reasonable.

The problem with this reasoning is that it misunderstands the way the court system works. Yes, each side presents the best possible evidence for its claim, so in that sense each side is biased, but what they present is supposed to be *evidence*. Evidence is a set of facts, also known as data, that support or establish a conclusion. Lies, fantasies, and wild speculations do not support anything and are not true evidence. If the court system were set up to hear extreme claims that are just wild speculations, there would be no reason to think the final outcome was just or true.[2] Expert witnesses play a special role in the process, because they are verified by the court to be especially reliable sources of evidence; in fact, judges must decide whether to accept someone as an expert witness before testimony is given. Claiming to be an expert witness while not attempting to be objective and reliable would amount to lying to the court, which is why it is unethical. It is also important to note that when an *engineer* behaves in this way, it casts the engineering profession in a bad light. The court will not be able to rely on engineers for expert testimony if they are known to communicate dishonestly and in a biased way when they serve as expert witnesses.

For these reasons, then, it is important for Sam to provide Edward with the most objective, well-supported evidence he can, so that Edward will be able to provide responsible expert testimony. Thus, Sam is ethically obligated to develop an honest, reasonable site plan. If Edward is unhappy about the final monetary value the plan gives to the land, he may pressure Sam to change the plan, but Sam would be unethical to give in to that pressure. This case is one of the many times when ethics requires courage and self-sacrifice.

Questions

1. What are the obligations of an expert witness?
2. What are the obligations that Edward and Sam have in this situation?

3. What specifically are the issues with the firm's receiving a percentage of the increase in compensation that results from Sam's testimony? If the court allowed the opposing side to raise this payment arrangement as an issue, could it damage the credibility of Edward and Sam?
4. What are potential consequences (professional and personal) if Sam greatly exaggerates the landowners' financial losses?
5. What is the best course of action for Sam while he maintains employment with Edward?

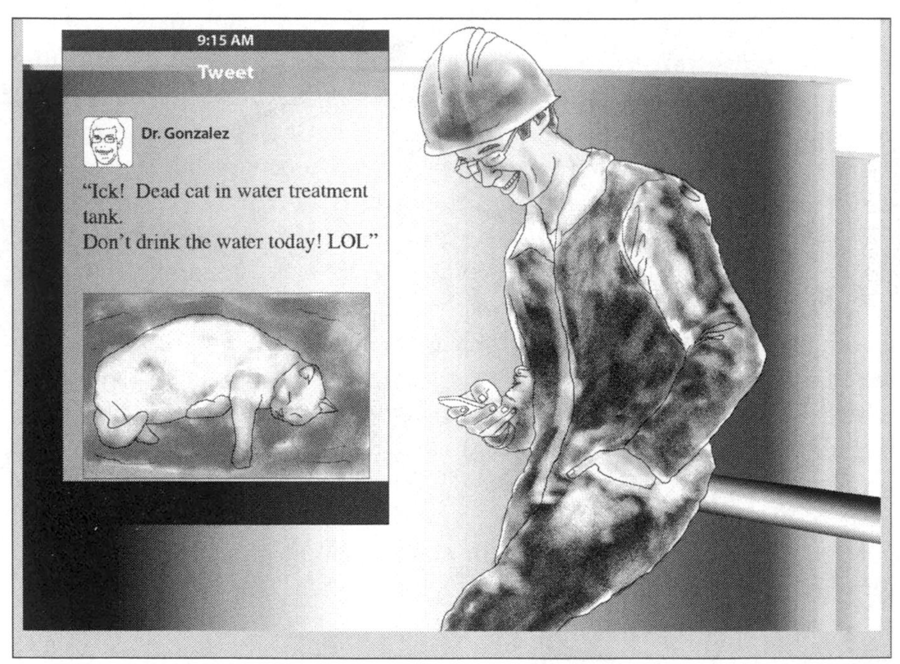

An engineer thinks he is telling a funny joke via social media.

Case 2. Hazardous Tweeting

Dr. Jim Gonzalez is a civil engineer and professor who is taking his class on a specially arranged tour of the city's water treatment facility. While the group is walking around the secured outdoor open-air water treatment tanks, they notice an animal floating in one of the large tanks. The supervisor scoops the animal out with a long pole and net, discovering it to be a drowned cat. Several class members express disgust, but the supervisor says, "Oh, it happens sometimes. Don't worry, there are plenty of treatment and disinfection steps following these tanks." Dr. Gonzalez thinks his friends would find the dead cat funny, so he pulls out his phone and writes a tweet: "Ick! Dead cat in water treatment tank. Don't drink the water today! LOL"

Discussion

Should Dr. Gonzalez post this tweet? This case appears simple at first. Communication via social media is very informal, so we are used to seeing it as the kind of communication that we leave up to individual character. Some people post pictures of their meals on Facebook, some post funny videos of cats, some post only impersonal links to news articles. None of these posting practices seem morally better or worse ways of using social media; they reflect different temperaments people have. Because many of these postings are purely for entertainment purposes, we generally don't expect them to be fact-checked and honest. So why should Dr. Gonzalez worry about what he tweets? His friends are probably used to his slightly twisted sense of humor, and they'll get a laugh out of his joke about contaminated water.

Unfortunately, things are not that simple. There are three features of Dr. Gonzalez's tweet that make it open to evaluation by Canon 3. In the first place, social media is *public*. It broadcasts speech across a group of people, and it does so in a way that can easily be recorded and passed on to more people, which is very different from a conversation around a dinner table. Because publicity is important for Canon 3, we should be careful to note that social media vary a lot in their publicity. Facebook is different from Twitter, and should be treated differently by Canon 3. Facebook enables one to control who sees one's posts, so there should be more freedom and more room for personal bias (that is, lack of objectivity) in one's speech on Facebook as long as one does in fact exercise control over the audience (it is very easy to forget to monitor security settings). Still, posts on Facebook are easily recorded and reproduced, so they are still public and require some care. Twitter is even more public. Anyone can follow another person's tweets and retweet his or her tweets with ease. Dr. Gonzalez is essentially speaking to the world when he tweets.

In the second place, Facebook, Twitter, and many other social media venues allow their users to include a profile, complete with educational credentials, employer, and other personal information. When people receive Dr. Gonzalez's tweets, they will be receiving them as words from an engineer and professor, not from a random person (unless Dr. Gonzalez is careful to conceal information on his profession), making it difficult for Dr. Gonzalez to wear a different hat when he is posting from his own accounts. It is almost as if he were speaking at a town hall meeting while wearing his company's official shirt and ID badge.

Third, Dr. Gonzalez is not merely communicating his personal thoughts about random topics. He is leading a tour of a treatment facility that he likely would not have been able to arrange if he weren't an engineer. He is seeing things the general public does not have easy access to. He is also tweeting about something that is within his realm of expertise, namely, water treatment procedures. His communication cannot help coming across as an expert judgment on water quality, regardless of his actual intentions. Furthermore, the content of what he is discussing is something of vital importance to every person: drinking water quality. People are especially likely to pay attention to what he is saying, and may be easily scared. Humans are hardwired to attend to threats to their own survival, and worries about contaminated water are always going to be easy to stoke.

Thus, our opinion is that Canon 3 would find Dr. Gonzalez's tweet to be irresponsible and unethical. He should save such jokes for private get-togethers with friends or perhaps an email sent from his personal email address to people he knows well.

Questions

1. What is the significance of the drowned animal in the water treatment tank?
2. Should this case even be considered an ethical dilemma? Shouldn't Dr. Gonzalez as an individual be able to tweet anything he wants?
3. Does Dr. Gonzalez have an ethical obligation to report this incident to the Public Works Director? Why or why not?
4. Does Dr. Gonzalez have an ethical obligation to ensure proper procedures are followed for such an incidence? Why or why not?
5. Should the public be informed through local media outlets of such an incidence? Why or why not?

Case 3. Environmental Test Results Only 1% Out of Compliance

Mia Samson was an environmental engineer with a few years of experience who worked at the consulting firm Across Engineering, LLC. One of the firm's major clients was a large manufacturing company, Supreme Devices Corp. Across Engineering was responsible for overseeing environmental tests for the various plant air stacks that Supreme Devices owned and operated. Across Engineering hired Ensure Air, LLC, to perform the actual tests. A test date was determined for a particular plant and the state was notified. Normally, representatives of the state would be present during air stack compliance testing, but on this particular occasion they were not. In attendance during the testing were Mia, Mickey Draper (a senior engineer from Across Engineering), an environmental compliance officer from Supreme Devices, and three Ensure Air staff members.

Ensure Air completed two of the three required air samplings and performed some preliminary analysis to see what the results were. It was determined that the air treatment stacks were 1% out of compliance. Supreme Devices would be fined heavily for not meeting the air quality standards. The individuals present discussed what to do next. It was agreed upon that Ensure Air would lie and tell the state that their testing equipment had malfunctioned and the testing would have to be performed at another time. This approach would allow Supreme Devices to modify the emissions treatment processes so that the standards would be met the next time the air stacks was tested.

Questions

1. What are the ethical issues involved with each party in this situation?
2. Did Mia and Mickey with Across Engineering meet the obligations of Canon 3? Why or why not?
3. How does the small amount of being out of compliance impact this situation?
4. What is the significance of Mickey's being at the testing compared to Mia's representing Across Engineering?
5. What is the best information from this situation for the state?

6. What is the best information from this situation for the public?
7. If Mia came to you for advice while in the middle of this situation, what would your advice be to her?

---

[1] Kant, I. (1963). *Lectures on ethics*, Trans., L. Infield. Hackett, Indianapolis, 226–227.

[2] In fact, there are extensive federal guidelines for evidence. See *Federal rules of evidence*. <http://federalevidence.com>.

Chapter Five

Act as a Faithful Agent

> **Canon 4.** *Engineers shall act in professional matters for each employer or client as faithful agents or trustees, and shall avoid conflicts of interest.*

Canon 4 focuses on the engineer's obligations to employers and clients rather than to the public. Such obligations can conflict with obligations to the public, so there is tension between this canon and those that are more focused on the public. This problem is not unique to engineers. Any person may feel conflicted between duties to friends, family, employers, or neighbors and duties to the larger public or the nation. A complete ethical theory, which should provide comprehensive guidance for our actions, includes guidelines for resolving such conflicts when they arise.

What is the source of the engineer's obligations to his or her employer and clients? The obligation can be seen as contractual, even if no formal contract is signed. When two people agree to an arrangement, each is morally bound by that agreement (as long as both parties understood the arrangement and freely agreed to it). Hiring an engineer and paying for the services of an engineer are both agreements of this kind. The employer or client and the engineer are presumed to be rational agents with interests that would be served by entering the agreement. Each agrees to behave in certain ways toward the other until the contract is dissolved.

What do engineers agree to do when they enter into contracts with employers and clients? Of course, there is a great variety of things they agree to do, depending on their specialization and the needs of the employer. But Canon 4 states that there is a further agreement engineers make that applies to all their contracts with employers and clients: to act as faithful agents or trustees of those employers or clients.

To understand the idea of a faithful agent or trustee, it is necessary to discuss the concept of interests. Many of our interests are conscious desires and values, which are what most people think of when they think of interests. For example, an individual may be interested in sports or have religious interests. But some of these conscious interests are universal: everyone desires food and water. What is important for our purposes, however, is that a person can have an interest in something without consciously recognizing it. In fact, the person may even deny the interest. For example, it is in the interests of children to be educated and to go to the dentist, but children frequently feel no conscious desire to do their homework or get a cavity filled. Why, then, do we say it is in their interest to do something even if they don't take an interest in it? It is because we believe that the activity is necessary for their well-being and will make their lives better. To connect this concept back to individuals' desires, we can say that the activity in question is one they *would* want to engage in if they had full information and full reasoning abilities. So, we can define interests as those things that contribute to a person's well-being or as those things that people would want if they were fully informed and rational.

It's now easier to understand what an agent and a trustee are. An agent is someone who carries out someone else's actions in accordance with that person's interests. For example, I might have a conscious desire to have my lawn mowed but no time to mow it myself. I hire an agent, a person who carries out the action that I want done. That person's movements are in part guided by my interests: the agent mows my lawn because I want it mowed, not because he or she has any personal interest in mowing my lawn. To persuade the agent to mow my lawn, I must make the action in his or her interest as well, for instance, by paying him or her. A person can be an agent for someone who is fully rational and capable of carrying out his or her own interests; the agent simply gets direction from the person for whom he or she is acting. Thus, trust is not a necessary part of this relationship. If the employer doesn't trust the agent, the employer can in principle supervise every movement of the agent and ensure the agent acts exactly as the employer wants.

A trustee is different. A trustee is someone who is entrusted to act in somebody else's interests. The connotation here is that the person whose interests are being served may not be competent to pursue those interests himself or herself. For example, parents are a kind of trustee for their children; they are not agents who carry out their children's orders. Parents are obligated to act in ways that serve their children's best long-term interests, even when the children are totally unaware that they have those interests.

Unfortunately, even adults are frequently in a position of relative incompetence with respect to carrying out their own interests, sometimes even with respect to knowing what their own interests are. Such incompetence arises because people rely so much on technologies that are very complex. I may have no idea that it is actually in my interests to have the wiring in my house designed in a particular way. Maybe I just have a really hard time understanding how electricity works. Fortunately for me, the people who wrote the electrical code and who constructed my house were acting as good trustees of my interests. That is, I could trust them to act in my interests even though I don't fully understand the actions that they took. My well-being is much better served in this case by trusting others to act for me than it would be if I wired my house myself. We are all in this position with respect to some technologies: nobody can fully understand them all in one lifetime.

Because engineers are professionals who engage in complex technical work, they act as both agents and trustees for their employers and clients. As agents, they must attend to the stated interests of those for whom they work. But as trustees they must sometimes act independently, using their own judgment about how to carry out a project. What Canon 4 requires is that even when acting independently, engineers must be faithful or true to the interests of the person for whom they are acting. In other words, Canon 4 states a background clause in all of an engineer's contracts with employers and clients, namely, the employer or client entrusts the engineer to act for the sake of the employer's or client's interests. In a sense, the requirement to be a faithful employee is an implicit clause in all employment contracts: no rational person would hire someone to act *against* the employer's interests or to act with no regard for the employer's interests. It is important for the engineering profession to make this duty explicit, because the nature of engineering work is so technical and employers or clients need to put a lot of trust into the engineer's judgment. They cannot give the engineer that trust unless they believe the engineer has their interests at heart.

We are now in a position to describe what a conflict of interest is. In the simplest possible terms, any interests that cannot both be satisfied can be said to conflict. We all have many conflicting interests in this sense; for example, a person may want to lose weight but also want to eat unhealthy food. But the phrase *conflict of interest* refers more specifically to an ethically important situation: namely, when a person has an ethical duty to act for the sake of certain interests but also has substantial interests that pull the person in a countervailing direction. Paradigm cases involve financial or personal interests that conflict with the duty to faithfully serve the interests of one's client or employer, but we have widened the definition of conflict of interest so that we can highlight the very similar ethical worries that arise in nonparadigm cases, such as the case of Georgio that follows. Without the broader definition, we might not notice the ethical problems in situations like Georgio's.

What exactly is the ethical problem with being in a conflict of interest situation? Is it wrong to have interests that go against one's ethical duty? No. These countervailing interests may not be unethical in themselves, but because they pull the person with a conflict of interest toward actions that are counter to the person's ethical duty, they increase the risk that the person will act unethically. For example, a judge has a duty to make impartial, fair decisions; this duty serves the interests of the public in general. If a judge were to preside over a case involving criminal charges against a beloved family member, the judge would also have a strong interest in not seeing his or her family member convicted. That interest is not morally wrong in itself; it stems from love of the family member. However, it gives the judge a strong incentive to push the criminal trial in a certain direction, and that is antithetical to the judge's duty to ensure the trial is fair.

One might think such conflicts are easily resolved: ethics requires us to put ethical duties over personal interests, so we can just decide to act ethically in cases where there is a conflict. Just as someone can choose to override a desire to eat chocolate cake for the sake of a stronger interest in losing weight, people should be able to override whatever interests they have that conflict with their higher ethical duty. Why can't a judge simply say, I will ignore the fact that the defendant is a family member and commit myself to running a fair trial?

It is in such situations where the problems of unconscious interests and implicit bias arise. It is very difficult for us to be aware of all the ways our interests are influencing our judgments. Psychological research shows us that interests can even influence perception. For example, even if a person is not consciously hungry, when the brain detects hunger it will start to direct attention to food-related signals, and pictures of food will even appear brighter than pictures of nonfood items.[1] In principle, any strong interest can affect our judgments in ways we are not aware of and cannot prevent. Whenever there is a conflict of interest, we must worry about whether the person is biased by the interest that is in conflict with his or her ethical duty. That interest may cause him or her to rationalize unethical behavior, fail to notice everything that duty requires, interpret situations in ways that serve his or her countervailing interest, or otherwise distort his or her perceptions.

This problem is of utmost importance for engineers, who are asked to serve as faithful *trustees* of others' interests. As we saw previously, being a faithful trustee requires making objective, independent judgments about what would best serve the interests of the trustee's clients and employers. For engineers, making such judgments will require clear-headed scientific judgment, and such judgments are not mechanical or rote. An engineer will need to sort through complicated facts and make decisions that require expertise. The right answer to an engineering problem is neither obvious nor predetermined, leaving ample room for personal bias to infiltrate the decision-making process and bias the results. It is therefore very important for engineers to eliminate or at least reduce their conflicts of interest. If they don't, they can't be sure (and nobody else can be sure) that they are truly operating as faithful trustees.

When it comes to conflicts of interest, Canon 4 is stringent. It says that engineers *shall* avoid such conflicts. That means engineers must do two things: recognize potential conflicts of interest and then avoid them. Recognizing a conflict of interest can be difficult because it's in the very nature of bias that we don't see our own biases. Working with a family member, for example, can seem idyllic. What could be better than spending all day with someone you love? It may be the nonfamily members you work with who will most clearly see the problems: the potential for favoritism, breaches of confidentiality, and other problems caused by the different relationships you have with family members and other nonfamily coworkers. You might see yourself as simply sharing a funny work story with your spouse, but it's a huge problem if you've shared confidential information about someone you both work with. The second case that follows will give another example of a conflict of interest that could be hard to recognize as a conflict at first.

Once an engineer has recognized a potential conflict of interest, what is that engineer required to do? Sometimes it can be enough simply to inform interested parties that there is a potential conflict. Why does that solve the problem? Because it allows those relying on the engineer's judgment to factor the conflict of interest into their own calculations. For example, if an employer knows that he or she has hired an engineer who also works at another company, the employer can decide whether to accept the risk that the engineer will share information gained at his or her company with the other employer. It may be a risk the employer is willing to take.

Merely informing others of a potential conflict of interest is not always the solution, however. If engineers accept gratuities from contractors, everyone directly involved in that exchange knows there's a potential conflict of interest. The people who don't know, and who have an interest in knowing about it, are other contractors and the public at large, because they will be affected if the engineer's engineering judgments end up being biased by the gratuity. What if engineers simply made the information public? They could broadcast the fact that engineers are now open to accepting gratuities for their engineering work. That would break down trust in the engineering profession. People count on engineers to make scientifically informed, objective judgments. If everyone knows engineers' judgments can be bought, then the judgments lose all value. Thus, the solution to this conflict of interest is not to accept gratuities in the first place, and to state that requirement in the Code of Ethics so that it's a norm people can count on.

Case 1. Expert Witness Offered a Cut of the Settlement

Theresa Greene, P.E., has served as an expert witness for dozens of plaintiffs over the last 30 years. Recently, she was presented with a unique opportunity. Attorneys for a class action lawsuit offered her 25% of her regular fee, plus 3% of any settlement or court ruling in favor of the plaintiff. Before making a decision on whether to accept this arrangement, Theresa studies the preliminary documents. In her technical opinion, the documented evidence strongly supports the claims of the lawsuit, and she would be able to provide extensive testimony in support of those claims. Because the settlement could be in the millions, her 3% share would far exceed what her regular fee would total. She's worried, though, because she has never seen this kind of payment offered before. Is there something shady about it?

Discussion

Sometimes it is difficult to tell when one might be in a conflict of interest situation. This case is not one of those times. (In fact, in some locations, this kind of fee agreement for expert witnesses is legally prohibited.) The conflict of interest facing Theresa is easy to spot. Conversely, if she serves as an expert witness, then both her role as an engineer and her role as a witness who has sworn to tell the truth command her to be as objective and honest as possible in her evaluation of the evidence. However, how much she will be paid for the hard work of evaluating the evidence largely depends on how the case comes out, and how the case comes out at least partly depends on how her evaluation of the evidence comes out. If she ends up spotting some data that support the defendant instead of the plaintiff (who hired her), she will have a strong financial incentive to ignore those data, downplay them, or search for more data to counter them.

If Theresa were to act on her financial incentive and examine the data in the hopes that they will support the plaintiff's case, she would be exercising the very opposite of good scientific reasoning. Looking for evidence in favor of one's hypothesis and overlooking countervailing evidence is known as confirmation bias, and it is one of the primary biases that the scientific method is meant to counteract. Deliberately acting on confirmation bias would be a way of renouncing one's status as a scientist. It's an action that lacks integrity.

Fortunately, the same features that make this conflict of interest so obvious also make it easier to counteract than a more hidden conflict of interest would be. One place the conflict could be mitigated is in the courtroom itself. The court system has some similarities to the practice of science. Both institutions are aiming at the truth. Science aims at the truth of how the universe works. Courts aim at the truth of who is responsible for what, and what the just penalty is. Because these two institutions have similar goals, they must contend with similar obstacles, such as biases, dishonesty, incomplete evidence, and other impediments to finding the truth. Scientists overcome these obstacles through meticulous collection of data, testing of opposing hypotheses, peer review, and other established methods. Courts take a different approach. They rely on an adversarial system of argument, on the grounds that if both sides have equal opportunity to present the best evidence in support of their side, the end result will be closest to the truth.

If Theresa takes this job, she will be subject to cross examination by the defendant. Any decent defense attorney will investigate how Theresa would be compensated, and would raise this issue during the trial. Once the issue is out in the open, the judge and jury members can think for themselves about whether Theresa's testimony might be biased. In fact, this compensation scheme may in fact backfire on the plaintiff by sowing doubt about Theresa's testimony.

Another place where the conflict of interest can be mitigated is in Theresa's own study of the evidence. She has the advantage of knowing she's facing a clear conflict of interest, so she can take extra steps to counteract her own biases. She can enlist the help of an engineering colleague to review her work on the case from the perspective of devil's advocate. She can also formulate alternative hypotheses about the evidence, including some that support the defendant's position, and test each hypothesis rigorously.

Thus, it's not necessarily unethical for Theresa to accept this compensation arrangement in jurisdictions where it is legally allowed. Notice, however, that the conflict of interest Theresa is facing here is faced by all expert witnesses who are paid by one side or the other, even if the compensation doesn't depend on the outcome of the case. The reason there's still a conflict of interest is owing to the contractual nature of hiring relationships that was discussed earlier in this chapter. Anyone who is hired and paid by someone is agreeing to serve the interests of whoever hired him or her. Plaintiffs and defendants have an interest in winning their case, and they hire expert witnesses to help them. When the expert is to be paid the same amount regardless of the outcome, the conflict is not as obvious, but it's still there. Thus, all expert witnesses, to truly uphold Canon 4 should follow Theresa's example and work hard to counteract their own biases. The adversarial court system will do some of the work, but it can't do it all. If juries decided to reject every expert witness's testimony on the grounds that expert witnesses are hired by one side or the other, the entire practice of giving expert testimony would collapse. To remain trustworthy, engineers need to protect their own integrity.

Questions

1. Do you think you would be able to be objective if you were an expert witness with Theresa's compensation method? How do you *know*, one way or the other?
2. Do you see any problems with a court system that relies on expert witnesses who are hired by each side? What if one side can't afford

to hire an expert? What if it's possible to find an expert to support any crazy conclusion?

3. How is Theresa's situation different from an attorney's accepting compensation on a contingency basis?
4. Conflict-of-interest problems often arise from lack of disclosure. If Theresa accepts this compensation arrangement, what is the best way for her to communicate this information?
5. What other conflict of interest situations are you aware of? How are they similar to and different from the situation Theresa is in?

Case 2. Confidential Reviewer Considers Breaking Confidentiality

Georgio Sanda works as a technical reviewer for the Department of Defense. He spends many hours studying the technical details of billion-dollar proposals. His current round of proposals is giving him a headache. Each of the proposals has considerable strengths, but each also has a few weaknesses. The weaknesses weren't obvious at first, but they became evident as he compared the proposals from different companies with each other. The proposals are highly confidential, but Georgio feels conflicted, because in his opinion, the best product would be a combination of the different concepts. Knowing how important it is for the new products developed through these contracts to be the best possible to ensure success of military action, Georgio is tempted to write comments on each proposal recommending that they add ideas that he has seen in the other proposals.

Discussion

Georgio's situation doesn't contain an obvious conflict of interest. There is a rule in place that he maintain the confidentiality of each proposal he reviews, and he is considering breaking that rule because he thinks it is interfering with achieving the best outcome. If there is a conflict here, it is between two approaches to ethical thought itself: rule-based deontology versus outcome-based consequentialism. In the context of an employment contract in engineering, it's clear that the deontological consideration is meant to win. A requirement of confidentiality is strict and not to be broken unless there is an imminent threat to life, limb, or property. So, in a sense, there is no ethical question for Georgio. He agreed to confidentiality and he must maintain it.

However, there is a deeper, more personal conflict, and Georgio needs to recognize that he is facing it. This conflict is between his duty to exercise his objective engineering judgment and his interest in seeing the best possible product developed. His employer, the Department of Defense, wants to fund the best projects and develop the best products. As a trustee of its interests, Georgio wants that as well. Presumably, he also has a personal stake in funding and developing the best projects or he would not have taken the job. He wants military action to be successful and efficient. He wants taxpayer money not to be wasted. And, of course, as an engineer he also wants the best products to be developed. As an engineer seeing a way to improve a process or finding a better solution to a problem, it's almost impossible not to say something. That's part of what it is to be a good engineer.

Yet Georgio is under a strict duty imposed by his employer to maintain the confidentiality of the proposals he is considering. His employer wants a good outcome but also insists on a confidential process. Why the insistence on confidentiality? Is it an arbitrary, unreasonable rule? Not at all. Companies invest a great deal of money in research and development, paying their engineers to create new designs. If the designs were not proprietary (i.e., owned as property by the company), there would be no hope of profiting from them, and then it would be irrational to invest in them to begin with, so the designs never would have been developed. If the Department of Defense cannot guarantee companies that the work they have put so much money into will remain theirs and at least potentially profitable, companies would hesitate to submit proposals to the Department of Defense. Thus, the promise of confidentiality must be kept, and Georgio cannot reveal proprietary details contained in one company's proposal to others.

Unfortunately, it is very difficult to tell what is proprietary and what is not. Georgio starts his review with his own expertise and ideas about what will work for the project. As he reviews the proposals, it is inevitable that he will gain a clearer idea about what would work best. Is that proprietary information? We all get new ideas just by thinking through an issue in more depth and talking to others. Sometimes seeing something that *doesn't* work can give us a good idea of what *will* work. Maybe Georgio already knows some things but doesn't remember them until his memory is triggered by something he sees in one of the proposals. Is it wrong now for him to use his improved sense of what would work for the project in reviewing each proposal? How can he refuse to use his own knowledge?

The same problem will arise whenever an engineer changes companies or works for multiple companies. We learn from each job we do, and in fact Canon 7 urges engineers to continually learn new things. Surely it can't be wrong to take some knowledge from one job to another job.

Companies, of course, have statements about what they consider to be proprietary, and the Department of Defense gives reviewers guidelines on what is to be kept confidential. But it's impossible to draw a precise line between proprietary and general (unowned) knowledge, just as it is impossible to tell exactly which parts of an idea you invented yourself and which came from others. All knowledge is connected, and it builds on earlier foundations.

Here is where the conflict of interest comes in. As Georgio reviews proposals, he needs to think about what he can say and what he can't, but there is no precise answer to that question, opening the door to the influence of bias. Because Georgio has an interest in seeing the best project developed, he will want to reveal things that a less-biased person would see as proprietary. He may tell himself he is not revealing too much or that he already knew something and didn't acquire the idea from one of the proposals. He needs to be on careful guard against that tendency in order to preserve companies' confidence and the confidential nature of the process. Because there is no way to completely avoid this conflict of interest, Georgio can only try to minimize it by reminding himself of the dangers.

Questions

1. What are Georgio's motives for wanting to communicate ways he thinks products could be improved?
2. What are the conflicting viewpoints involved with this case?
3. Why is confidentiality important with regard to proprietary information?
4. Do any of the other canons provide guidance for how Georgio should proceed?
5. Do any canons conflict with each other?

A client attempts to influence a consulting engineer and increase competitive edge.

Case 3. Competitors Seek an Advantage

Henry Short works at a consulting firm and has numerous government and rural water districts as long-standing clients. Each client is aware that Henry does engineering for their competitors. The city of Smithville is expanding to the east, and two rural water districts that are Henry's clients are hoping to expand their services into that area. During a recent meeting with leadership of one of the water districts, Timber Creek Water District, Henry was asked point blank to propose a more difficult and costly pipeline route for their competitor, Prairie Creek Water District.

Questions

1. Why would Timber Creek make this request of Henry?
2. What key issues in this case constitute a conflict of interest for an engineer?

3. What aspects of this case may pose a conflict of interest for Henry?
4. What are Henry's options?
5. How would you advise Henry if he asked for your advice on this?

---

[1] McClelland, D., and Atkinson, J. (1948). "The projective expression of needs: I. The effect of different intensities of the hunger drive on perception." *J. Psych.*, 25(2), 205–222.

Chapter Six

Reputation by Merit

Canon 5. *Engineers shall build their professional reputation on the merit of their services and shall not compete unfairly with others.*

Why do we use the verb *to build* when we are talking about a person's reputation? Wikipedia defines *reputation* as "a component of identity as defined by others," so how do *we* build our reputation if, by definition, our reputation is *defined by others*? A quick online search reveals that there are companies out there that purport to give you the tools to take control of and manage your reputation![1] The fifth ASCE canon states that engineers shall build their reputation (presumably a *good* reputation), but then adds that they must do so "on the merits of their service." What does that mean? And if that weren't enough, the canon continues, "and shall not compete unfairly with others." Why must engineers compete fairly? What, at any rate, does *fairly* mean in this context? Decades ago the National Society of Professional Engineers (NSPE) in its code of ethics prohibited participating in competitive bidding. The concept was that engineers should be selected on their qualifications alone, and not on a lowest bid criterion. In 1977 the courts (USA v. NSPE)[2] decided it was illegal for the NSPE to prohibit engineers from participating in competitive bids. The engineering profession, however, still strongly advocates a qualifications-based selection of engineers for projects.

As we mentioned in our introduction in Chapter One, Aristotle thought that we "are what we do habitually,"[3] so our habits form an integral part of the kind of person we are. Because our habits are often noticed by people around us, our reputation works in a similar way. When we say something like, "Gary Washington has a reputation for being late," we mean that our impression of Gary includes an often-noticed propensity to be late, say, to company meetings. Gary could have earned that reputation only one way: by being late to many company meetings. Of course, Gary cannot directly control what others think of him, but he can control the actions he takes that shape what others think of him. So Gary can rebuild his reputation only by changing his habits and arriving on time, not simply to a meeting or two, but to every meeting from now on. In due time, people will start to say, "Gary? He's *always* here on time!" and might think it unusual if the meeting is about to start and he's not present. That is how reputation works. Yes, it's what others think of us, but they think those things of us because of what we do.

A professional reputation, then, refers to the collective impressions that people have about our professional habits, including strengths and weaknesses, technical competence, and demeanor. Canon 5 requires that we build our reputations only on the basis of our work. When we produce solid results over and over again, we build a good reputation. Engineers, unlike, for instance, lawyers, cannot use billboards, television ads, or other such techniques to bolster a positive image: we must rely solely on our work and results.

And we must compete fairly. What does *fairly* mean in this case? It means that when competing for a contract, for example, we must make available our best samples of our work and let the chips fall where they may. In other words, if another firm presents better work, has better qualified engineers, or has other advantages to offer, it only makes sense that they are awarded the contract. Engineers must not embellish their credentials, falsify their resumes, or misrepresent their level or variety of expertise.

A harder question to answer is this: Why should an engineer compete fairly? We sometimes hear the expression "all is fair in love and war." Setting aside the obvious problem with the comparison between war and engineering, it seems as though we do need to answer the question. Is it justified to do whatever necessary to get the contract?

If one is compelled to answer yes, then we are back to the question, do the ends justify the means? As we saw in our introduction in Chapter One, this approach is clearly utilitarian (consequence based). And the answer to the question is, maybe sometimes the ends justify the means, but then something has to justify the ends. In this case, if the end is being awarded the contract, then, no, competing unfairly is not a justified means. Why? Because, as principle-based ethics would show, a rule of action that allowed engineers to embellish their credentials simply to be awarded a contract would never be accepted as a universal principle. If everyone embellished their credentials, no one would believe such credentials in the first place. A rule that can be universalized is this one: "Engineers ought to be truthful about their merits when competing for a contract."

Let us not brush aside consequence-based ethics too quickly, however. Under the proper analysis, we can use a utilitarian approach to disapprove of the means (embellishing a resume) in order to achieve a desired end (being awarded the contract). All we have to do is ask which result—or end—is the one that matters: that a particular engineer is awarded the contract or that a given project has the best-qualified engineer leading it? Canon 1 reminds us that "engineers shall hold paramount the safety, health, and welfare of the public," and competing unfairly would jeopardize the safety, health, and welfare of the public. The more unqualified or underqualified engineers are working on a project, the greater the risk to the public.

Canon 5, then, leans on character-based ethics for the first part (regarding reputation) and largely on principle-based ethics for the second part (regarding competing fairly). Even though reputation is defined as what others think of someone and his or her work, the only thing engineers have under their control is their own actions, and the pattern of their actions over the long run is what Aristotle meant by character. This is why we say we "build" our reputation: we influence what others think of us purely by how we represent ourselves, through the merits of our service and our professional habits.

Before we turn to our case studies, there is one other question that needs to be addressed: How do we know when our actions result in an unfair advantage? Some individuals claim anything is fair when competing for business, and certainly we recognize how fierce competition can be. To make things more complicated, outside of the business world, the arenas of competition that we are all familiar with, such as sports and games, have very different rules of fair competition. In baseball, it's not unfair to bring in your star fastball pitcher to face just one batter, but it is unfair for that pitcher to put pine tar on the baseball. In poker, it's not unfair to read your opponents' facial expressions to get a clue as to the cards they're holding, but it is unfair to install a secret mirror so you can see their cards. Yet in baseball it's fair if a third base coach sees the catcher's pitch signals and communicates them to the batter, but it wouldn't be okay for a poker player to have a third party try to steal glances at other players' cards and communicate the results. Why all the different rules? Because the rules create the field of competition and define what the game is about. If baseball allowed pitchers to custom-coat their balls, it would be a very different kind of game, but as long as the rule was known and applied fairly, it would still be a fair game. If poker allowed players to see each other's hands, it would be a different game but still a fair one if the rule was known and applied in a fair way. The problem arises when the rules are set but secretly not followed by some, which undermines the purpose of the game.

Engineering is, of course, not a game. But it has a purpose and rules of fair play like any other competitive domain. The purpose of the rules is to ensure that high-quality, effective engineering projects are completed in an efficient and ethical manner. Once the rules are known, secretly going around those rules is defined as unfair. Here are some examples of what we think constitutes an unfair advantage:

- An engineering firm owner using his influence and access to confidential information as a county commissioner to obtain contracts for his company.
- An engineering firm hiring a new employee at an inflated wage, strictly because that employee's family owns a big land development company that regularly hires consulting firms for projects.
- An engineering firm "accidentally" leaving a phony project proposal where the competition can find it so that the competition would be misled.
- An engineering firm reading confidential information created by another firm.
- An engineering firm paying informants for inside information about the bids of competitors.
- An engineer creating an artificially low bid on a design/build project with the expectation that there will be many change orders that would greatly improve the profitability of the project.
- An engineer falsifying qualifications to win a request-for-qualifications competition.
- An engineering firm claiming a lead role in projects for advertising purposes when in fact the firm's involvement was minor.
- An engineering firm exaggerating the role of minorities or women in the firm to gain minority-targeted contracts.

As you can see, these are examples of people or companies acting in such a way that results in an uneven playing field, not because they have better qualifications (i.e., the merits of their services), but because they take advantage of a loophole, concoct and disseminate false information, or provide false or misleading information. Let us now turn to our case studies for examples.

Case 1. Generous Competitor

The small town of Twin Forks recently awarded another sewer lift station project to a competitor of Alfred's. Alfred Chen, P.E., had submitted bids on many small projects around Twin Forks but had not been awarded any of them. His competitor, Richard Mays, P.E., had received all the projects. Alfred wanted to learn why he was not awarded even one of them. After all, his credentials and reputation were comparable to Richard's; in some cases, Alfred thought he even had an edge over Richard. One day, Alfred went through the notes of many of the city council meetings and saw several mentions of Richard making financial contributions to city park funds. He looked up some of the dates of the previous projects and determined that Richard had consistently made contributions to the park fund after the completion of each project. It seemed to Alfred that the Twin Forks city council was awarding projects to Richard partly because the council knew he would make a financial contribution to the city's park fund. The projects Richard was awarded were largely funded by state coffers, so if an engineer increased a bid to provide resources back to the city, the city might think it was a win–win situation. Alfred was now faced with determining what to do next.

Discussion

The question here is whether Richard is doing something unethical by donating money to the city parks. Do Richard's actions constitute unfair competition? There may not be anything illegal about Richard's donations, but are they consistent with Canon 5 of ASCE's Code of Ethics? And can—or should—Alfred do anything about it?

The central issue, then, is the donations. Are they—or can they be construed as—creating an unfair advantage? It seems that a reasonable, disinterested third party would say that they are. Even though the pattern seems to be that the donations come after the completion of the projects, it is not hard to imagine that the various parties involved know what is occurring: Richard is awarded the project, and a few months later, the city park receives a nice donation from Richard. It's also easy to assume that those who choose who is awarded the contract know that Richard is the one who makes donations after construction is completed. More to the point, this advantage has nothing to do with the merit of Richard's services. Remember, in terms of merit alone, Alfred is at least as good as Richard, so the donations could be perceived to be influencing the decision makers.

Conversely, Alfred doesn't know for certain that Richard is competing unfairly. It is possible that the timing of Richard's donations is coincidental and that Alfred has simply had bad luck in securing contracts. Perhaps Alfred is doing a poor job of presenting himself; maybe his qualifications aren't highlighted well enough or perhaps his bid submissions look less professional than Richard's.

Does Alfred need certainty about unethical behavior before he speaks up? No. If a reasonable, disinterested third party would suspect that there is a problem, there is adequate reason to raise concerns and initiate a discussion and possibly an investigation. A profession can maintain a strong ethical culture only if people speak up when they have concerns. A profession is not a police force that requires probable cause before investigating, and ethical principles are not laws that can be enforced only through punishment. Ethical principles are ideals for which we can all strive, and the profession has a right and responsibility to make sure those ideals are being honored. An atmosphere of open discussion and willingness to raise questions helps keep the ideals strong.

In a case like this, Alfred should speak up. He could report Richard to the city board or report the matter to the Ethics Hotline at ASCE or to the State Board of Engineers, or Alfred could even discuss the issue with Richard directly. The facts of this case show that Richard may be competing unfairly and Alfred shouldn't have to take it quietly.

Questions

1. What do you think about engineers providing contributions the way Richard did in this case? When are these gifts acceptable or unacceptable?
2. Why is Richard providing these park fund contributions? Are Richard's actions consistent with the ASCE Code of Ethics?
3. Would it be okay for Alfred to include a larger, hidden contribution to the park fund in a future proposal for a project and to be sure the government staff knew he was planning on making a substantial contribution after completion of the project (i.e., larger than what Richard had been giving)?
4. If Alfred believes giving donations to the park fund is unethical conduct by Richard, what are his options?

An engineer overhears a conversation about bids on a project.

Case 2. Bid Leak

Anthony Long, P.E., had recently submitted a bid for a commercial land development project along Highway 19. He stopped by the county Geographic Information Systems (GIS) office to get some additional information about another project and then he went to get a cup of coffee. As he approached the breakroom, Anthony overheard a conversation. It was one of his competitors, Bob Gifford, P.E., and the developer, David Ford, speaking about the same Highway 19 development project. Bob said, "So, Anthony's team always puts together a competitive bid, don't they? Say, if I were to submit a bid for, oh, $15.1 million, do you think that would be in the ballpark?" David paused and then replied, "$15.1 million sounds a little high for that project." Anthony was furious, and he considered whether he should confront the two individuals or leave and report the situation.

Discussion

Bidding on construction projects is an integral part of the life cycle of an engineering project. When a project is advertised, basic information about the scope of the project is released so that engineering construction companies can put together their best-guess estimates as to how they would complete the project, ending, of course, with a price. These bids must be generated independently by each company competing for the project. If one company knew the bottom-line price of a competitor's bid but the competitor did not have the same information, the company with the information could be in a position to submit a more attractive bid. The winner would always be the company with access to others' supposedly confidential information; it would have an unfair advantage with the result being that it could be chosen over a company that is better qualified, which is why the construction bidding process is rife with rules and measures to ensure impartiality and anonymity.

This case study presents us with another clear example of an uneven playing field. The conversation between Bob and David is inappropriate because—however indirectly and circuitously—David is releasing information to Bob about a competitor's bid. Bob can then go back to his office and adjust his numbers, thus increasing his chances of being awarded the project. Bob, therefore, competed unfairly in this case, because he did not allow the merits of his services to speak for themselves.

It would be possible to set up the rules of bidding for engineering projects so that every bidder knew everyone else's bids, but that would be very inefficient. Each firm would need to adjust its bids after seeing the others' bids, and then those firms would adjust their bids until everyone reached the same bottom-line price, with profits cut to the bone. After all, making a tiny profit is better than never being awarded any contract. Engineering firms could not flourish under such a system, so a confidential system is set up when bids are solicited, and once it has been set up, breaking the rules of anonymity constitutes cheating.

Anthony has every right to be angry. And he has the right—and perhaps the obligation—to confront the two men about what he overheard. Given the potential adverse effect that a personal confrontation might have, an alternative would be for Anthony to report the matter to ASCE's Ethics Hotline, or, if available, to an ombudsman at the company that advertised the project.

Questions

1. What are the principles behind sealed bids?
2. Why was Bob asking David about this project?

3. Should Bob be asking David about project details? What kinds of questions are acceptable?
4. What implications does Bob and David's conversation have for Anthony?
5. What specifically is the ethical misconduct?
6. What should Anthony do?

Case 3. Changing Dam Safety Standards

Sheila Inger's engineering firm was selected to oversee the construction of a sizable dam. The design process had taken years to accomplish because of the complexity of the situation and a lengthy approval process. The start of construction had been delayed because of the lack of funding. Construction had finally started, and Sheila was asked to perform a formal downstream hazard analysis. The hazard classification criteria had changed since the dam was designed, but the design was expected to meet the new requirements. Upon determining the average daily vehicle count and completing her analysis, Sheila determined that in fact the hazard classification had changed from a low hazard to a significant hazard. This change was caused by the state's new criteria. Moreover, the average vehicle daily count on a nearby downstream highway went from 481 to 503 since the original design was done. An average daily vehicle count of 500 makes the road a moderate-volume road, which affects the dam's hazard classification. Sheila knew it would be a huge project to redesign the dam to meet the new hazard requirements. She debated with herself what to do. Sheila knew her firm would be likely to receive the new design contract; however, it seemed like a waste of resources to redesign the entire dam because traffic just barely met the moderate-volume level.

Questions

1. Is it a big deal to redesign a dam for a more stringent hazard classification?
2. How does the lack of resources impact Sheila's situation?
3. Is the new average daily vehicle count of 503 significantly different from the previous 481?

4. Sheila's firm would likely receive a lot of new work because of the higher hazard classification. So why is this an ethical dilemma for her?
5. What do you think the client wants Sheila to do?
6. What would you do if you were Sheila?

---

[1] See, for example, http://www.reputationmanagementconsultants.com/.
[2] United States of America v. National Society of Professional Engineers. 1977. 555F.2d 978. 181 U.S.App.D.C.41, 1977-1 Trade Cases 61, 317. Open Jurist online resources. <http://openjurist.org/555/f2d/978/united-states-v-national-society-of-professional-engineers>.
[3] *Nicomachean Ethics*, at marginal pagination 1103b05.

Chapter Seven

Uphold Professional Honor

Canon 6. *Engineers shall act in such a manner as to uphold and enhance the honor, integrity, and dignity of the engineering profession and shall act with zero-tolerance for bribery, fraud, and corruption.*

Canon 6 has two parts (or imperatives). The first is to "uphold and enhance the honor, integrity, and dignity of the engineering profession" and the second is to "act with zero-tolerance for bribery, fraud, and corruption." As you can see, there are a number of terms we need to define at the outset: *honor, integrity,* and *dignity* for the first part and *bribery, fraud,* and *corruption* for the second. Let us start with the first three terms, which mean somewhat similar things.

The word *honor* conjures up images like that of a soldier accepting a medal (such as the Medal of *Honor*), a bailiff announcing the judge entering a courtroom ("All rise! The *Honorable* Christine Davis presiding . . . "), and perhaps the opening words of a speech delivered by a guest at a prestigious conference ("First I want to say how *honored* I am to be here . . . "). Honor, then, refers to an acknowledgment of *credit, reputation,* or *good name* of a thing, person, or concept. When we say we ought to honor a promise, for example, we are saying that we should give the concept of promise keeping its due credit or importance. To honor a profession, therefore, means to treat it with the value that it deserves.

The engineering profession enjoys a high level of prestige. To uphold and enhance the honor of the engineering profession means that engineers should act in such a way as to keep the reputation of the engineering profession where it is (uphold) and to improve (enhance) it whenever possible. When engineers fail to do their jobs—when they are caught compromising their values, for instance—they may suffer personal consequences, yes, but a greater damage is likely: they may be hurting the prestige of the engineering profession as a whole. To honor the engineering profession, therefore, is to act only in ways that keep or improve its image and reputation.

To behave with *integrity*, a term rooted in the Latin word for "*whole*," means to adhere to one's moral principles, to stay true to something. A person of integrity earns a reputation of always acting in accordance with proper moral choices. Aristotle would argue that integrity is a virtue that is developed by the habit of acting morally over a long period, a career perhaps. Again, to uphold and enhance the integrity of a profession is to always, always act in accordance with the ethical standards of that profession. In our case, we can say that this part of Canon 6 is essentially stating that we must develop a habit of following all the canons of the ASCE Code of Ethics.

Last, *dignity* stems from the Latin word for *worth*, so in a similar fashion as honor, it refers to the acknowledgment of the *worth* or *value* of something. To uphold and enhance the dignity of the engineering profession is to act in such a way as to keep or improve its worth. Again, getting caught bending the rules in an engineering contract may get the engineer in question fired, but it also harms the dignity of the engineering profession as a whole.

Let us define the next three terms, the ones related to the second imperative: *bribery*, *fraud*, and *corruption*. A fourth term, *extortion*, will be added to this list for reasons that will become clear.

A *bribe* is a payment made to an individual or organization apart from the money exchanged for products or services rendered, given for the sole purpose of gaining an advantage. As we saw in Chapter 6, an engineer must compete fairly; a bribe is, by design, an attempt to gain an unfair advantage over others. For example, if an engineer, Corey Ingersoll, submits a proposal for a project and knows that a competitor has a superior product, he might be tempted to offer the person selecting the winning bid, Frank Gupta, a bribe. Frank may know that by selecting Corey's bid, the firm would not be receiving the best product available, but that's okay—he's getting a new hot tub. As we saw when we evaluated Canon 5, a bribe like this is not an example of competing fairly.

It might prove useful here to define an additional term, one that can sometimes be confused with bribery, but that also carries its share of difficulties: *extortion*. Extortion is a payment demanded by a person or entity for services that the customer is entitled to but that go above and beyond the ordinary, agreed-upon transaction. Interestingly, the Latin origin of the word "extortion" (*ex*,- of, and *torqu re*, twisting or wringing) conjures up just the right image: an attempt to squeeze some more money out of a person because of the situation.

Suppose that a driver, Matt O'Connor, needs to transport some construction materials to a project site in a neighboring country. The taxes, fees, and tariffs have all been taken care of and the transfer is perfectly legal, but when the truck reaches the border, a customs agent, Jeff Beaman, orders Matt to stop. Matt shows the agent all the relevant paperwork, but the agent stalls. After a couple of hours of waiting and frustrating bickering, Jeff offers to let the truck through—at a price. Jeff calls it a transfer fee. The industry calls it a grease payment. The proper term is *extortion*.

Notice that if this transfer fee payment is made, it is not a bribe. To make the difference more clear, let's change the situation a bit. Suppose Matt reaches the border and sees a huge line of trucks waiting to go through. Nothing amiss here; it's just the normal, expected delay that sometimes happens when a truck is crossing international borders. Jeff walks to the truck and informs Matt that he'll have to wait about three hours before it's his turn. Now, if Matt offers Jeff money to expedite the process (i.e., grease the skids, in effect to let Matt cut in front of everyone else in line), that is a bribe. The distinction hinges on whether the payment is intended for a product or service for which the customer is already entitled (extortion) or enabling a state of affairs for which the person is not entitled (bribe).

Fraud is deceit, trickery, or breach of confidence perpetrated to gain a financial advantage or an unfair or dishonest advantage. Essentially fraud means to deceive, as in intentionally saying something that is false.

Suppose there is a U.S. Army Corps of Engineers project that carries a Buy American clause. If Clarence Hobbs submits a bid that states he is complying with the clause when in reality he is not, he would be committing an act of fraud. If the Corps of Engineers selects Clarence's bid because it had, for instance, the best price, the Corps would not be receiving the products the contract requires.

Corruption is a more general term meaning *moral perversion* or *depravity*, in effect an umbrella word for practices such as bribery, extortion, and fraud. Measuring the overall economic impact of corruption is remarkably difficult, in part because most of the literature on the subject, at least until recently, focused on the perception of the impact of corrupt practices by various parties affected.[1,2] A senior economist at the World Bank, however, states, "A low-end estimate suggests that the financial costs of corruption in infrastructure investment and maintenance alone in developing countries might equal $18 billion a year," and adds that "estimates regarding the cost of corruption in infrastructure suggest that 5 to 20% of construction costs are being lost to bribe payments."[3] Clearly, then, corruption is a problem that costs real money, time, effort, and even lives, so this second part of Canon 6 carries the imperative for engineers to fight corruption to the utmost of their abilities.[4]

The meaning of the canon seems straightforward: the first part brings up the need for the engineer to do the right thing, whereas the second part states that the engineer should refrain from doing bad things. Simple, right? Let's take a closer look.

On television, "Larry, the Cable Guy" often encourages people to "git-r-done!"[5] Does this mean that we should get the job done no matter what? Of course not. But that is what can be at stake here: sometimes getting things done comes (or at least seems to come) with a need to bend the rules or flat out ignore them. Is it really feasible to have zero-tolerance for bribery, extortion, fraud, and corruption in all aspects of an engineer's job, even for multimillion dollar projects? What if the project in question is in a foreign country? What if the project would not be possible without bribing at least a few corrupt officials? Isn't a small grease payment worth it when the result is a highly lucrative project?

Especially when the conversation involves foreign countries, there is a tendency to revert to the cultural relativist's argument, where statements such as "that's just the way it's done here," "it's the cost of doing business," and "it's a necessary evil" come into play. It becomes easy to blame those "other guys" for bribes and grease payments needed to keep a project afloat. Let's not kid ourselves, either: there's plenty of bribery, extortion, fraud, and corruption to go around anywhere in the world.

A common attempt to justify or rationalize such actions involves changing the name to make them more morally palatable. People taking bribes become "project advisors" or "government liaisons" so that their salaries or fees appear as bona fide project expenses. Extortion or grease payments, as in the example of the truck driver, are logged and therefore somewhat legitimized as a transfer fee. Such euphemisms, however, cloud the issue. Corruption is morally reprehensible—whatever shape it takes and whatever you wish to call it. We discuss the many forms and names that corruption takes in more detail when we tackle our case studies.

Let us recall the concept of cultural relativism, brought to light in our introduction in Chapter One. Cultural relativism, that is, the conclusion that there are no universal moral standards, rests on faulty logic.[6]

The premises of this argument have to do with what people believe to be moral, yet the conclusion refers to what is moral (i.e., nothing is absolutely moral). This reasoning is fallacious: we can't go from what people believe to be the case to what is the case. Even if we took the moral pluralists' position, where various moral theories might be right in their own context, we would find that all three major moral theories would be consistent with a decision to avoid bribes, extortion, and fraud.

Someone might object: Wouldn't it be possible that a consequence-based argument justifies a bribe? What if a small bribe leads to enormous profits? The greater good would be achieved by bribing the inspector, wouldn't it? Recall, however, that consequence-based theories ought to consider the results of our actions not in the immediate sense but in the long run. Greatest good for the greatest number doesn't mean greatest good for the company that employs the engineer.

This change of perspective would force us to think beyond the bribe, which we might consider a minor expense, to the consequence of the practice of bribing. The same can be said for extortion. As we noted previously, when an estimated 5% to 20% of construction costs end up in the hands of corrupt inspectors, government officials, or convince an engineer or designer to cut corners, the consequentialist math tells us that a world without corruption mechanisms is better than the world with them. The same goes for when bribes are used to obtain an inferior product. Yes, maybe the company making the bribe spent just a few dollars to save a lot, but the result is inferior products—thinner concrete, weaker gauge steel, or other unacceptable material—which in turn result in higher maintenance costs, reduction in safety, higher impacts on the environment, and so on.

Let us now return to the issue of *zero tolerance*. Suppose that while working on a project in a remote area of the world, you encounter a mail carrier who demands an additional fee to deliver business mail to you. This mail carrier doesn't get paid well, and your jobsite is quite out of the way. It's that simple: if you don't pay him the fee, you don't receive the mail necessary to manage your project. As defined, this fee is really a grease payment, a textbook example of extortion. Is this the same as a government official demanding a kickback to approve an engineering project? Well, it is and it isn't. Both are examples of extortion, yes. But we can see how an engineer will not be able to solve the world's problems by refusing to pay the mail carrier's fee. All that engineer will accomplish is to be out of mail. A government official, however, might be a different story.

He or she represents a country, and an engineer represents the company doing business in that country. There are a number of systems and processes that the engineer can use to remedy this situation.

Are there practices common in our own culture that might be construed as being similar to bribes? Is tipping a bartender a bribe? Most people might say that tips are not bribes; they are an expression of gratitude (hence the more formal term gratuities). We can, however, imagine that even in this scenario, bribes are possible: maybe when we tip our bartender generously we get a stronger drink (i.e., the bartender benefits personally, while the bar owner loses money on the drinks with more alcohol).

What about political lobbying? Do lobbyists essentially bribe politicians when they take them out to dinner or on an all-expenses-paid trip? Is it possible that lobbying is a legal form of bribery? The industry that is employing lobbyists might see their services as a legitimate and necessary expense to maintain a profitable business, but does this expense result in an unfair advantage? When do expressions of gratitude result in expectations of favoritism?

No one is saying that these are easy questions to answer. Does the engineer tough it out without mail and report the mail carrier to the post office for dereliction of duty? Or does the engineer acquiesce to the mail carrier's easily manageable demands? Does the politician accept the trip or not? These are hard questions. Canon 6, however, calls for engineers to have "zero-tolerance for bribery." There is no distinction made between mail carrier-like grease payments and government official-style extortion. A principle-based approach would condemn both equally, whereas a consequence-based approach might be invoked to argue that paying off the mail carrier may be necessary—or even inevitable—to produce a much larger benefit, although, as we have argued, when considered in the long run, even consequentialist arguments reject such corrupt practices.

There is one more way to answer cynics who argue that zero tolerance for corruption is an impossible standard to uphold: ask them to provide an alternative. In other words, who in their right mind would argue, "If zero tolerance is unrealistic, then, okay, let's go with 10% tolerance"? What percentage would be realistic? 10%? 5%? 1%? How would we even quantify that we'll allow for 5% tolerance? We can readily see that this answer is a nonstarter. The standard must be zero tolerance. This standard is not too much to ask. We are, however, still left with the option to legitimately debate the moral status of our actions: is this a bribe or is it a mere expression of gratitude? Does a certain payment result in an unfair advantage, or is it a morally defensible cost of doing business?

Case 1. Negotiating Truth

Burnell Wainwright is a military engineer in the U.S. Army who is overseeing construction in Afghanistan. The funding is coming from the international community to help the country rebuild critical infrastructure after a devastating war. A small local company, QualSpec, was placed under contract to provide construction inspectors to help Burnell stay informed on construction progress. Construction inspections are valuable for any sizable projects, but inspections were vital to these new facilities so that acceptable quality was ensured. In this particular environment, construction without inspection would have created many facilities that were not functional or would last just a few years when the life span should be near 50 years.

Burnell and QualSpec worked together to select the inspectors and many local engineers were hired. The hourly wages being paid to these inspectors were competitive for the region, but quite low in comparison with similar positions in the United States. In fact, QualSpec was receiving about $60 per billable hour, and Burnell thought the inspectors would be paid about $15 per hour. QualSpec was actually paying the inspectors between $1 and $2 per hour and had a line of local engineers at the door desiring this work at that extremely low wage.

After completion of the first week of work, the construction inspectors submitted their hours on a standardized timesheet. Each inspector submitted about 140 hours of work for the first week. Burnell couldn't imagine why each inspector would report 20 hours per day for all 7 days. It must be some simple error to be wrong by a factor of about 3. So he asked the inspectors to recalculate and report again. They reported 120 hours. Burnell was shocked, so he called a meeting with all the inspectors. He informed them he knew there was no way they each worked 120 hours in the first week. The inspectors huddled together for a bit and then replied, "Okay, we accept we won't be paid for 120 hours. So, what's your offer?" They explained to Burnell that they were only being compensated $1 to $2 per hour and could barely survive on that wage. It was apparent they were accustomed to negotiating virtually all business arrangements and they viewed this job no differently, so they were ready to engage in a standard business negotiation.

Discussion

Burnell's situation is complicated. The first morally salient feature of this case study is that the construction inspectors are essentially lying on their timesheets. It's that simple: they indicated on the form that they each worked 140 hours that first week (down to 120 before they asked Burnell for his counteroffer), whereas in reality, they probably worked half that much. On that account alone, the decision should be straightforward: figure out a way to estimate the actual hours worked and pay them that amount.

In this case, however, the complexity of the situation deserves closer scrutiny. First, we must ask ourselves if the construction inspectors are, in fact, requesting (or requiring) something that they are entitled to, or, for that matter, if they themselves think they are doing something fraudulent. A cultural relativist would quickly point out that negotiating is commonplace in many foreign countries, so they are simply doing what comes naturally to them. As we pointed out previously, cultural relativism ought not to be used to excuse or justify fraud or to tolerate corrupt practices, however ubiquitous or commonplace it might be in that culture. We will return to that concern, but first let us explore a few more variables.

Afghanistan is not a typical location for an engineering/construction project. It is a war-torn country. The U.S. Army is there to help the country return to a sense of normalcy, peace, and independence. The U.S. government has deemed it a worthy cause to invest in rebuilding the infrastructure of the country (much of which was destroyed by United States-led airstrikes), and Burnell is in the difficult position to manage the construction projects in his region.

In evaluating the situation at hand, it would seem to be patently reasonable for Burnell to ask the following questions: If he were to pay the inspectors exactly what the contract requires, based on the actual (not reported) hours worked, would it amount to reasonable compensation? Clearly, we would have to articulate some standard for what "reasonable compensation" would mean in this case, but what if the answer to this question were a resounding "not even close!" under even the most conservative standards? The fact that these inspectors must work under extremely difficult and dangerous conditions (i.e., they could be shot at because they are associated with the United States' rebuilding efforts) should justify increasing their compensation significantly, right? What if Burnell knew that construction inspection leaders of adjacent regions were paying the overreported hours without raising a fuss? Would the lower compensation result in substandard performance from the inspectors? Maybe they would be unable to perform their duties with the downward-adjusted pay.

It may be worth mentioning what is not going on in this case, just to highlight conditions that would significantly change our moral evaluation of Burnell's situation. The inspectors, for example, did not offer Burnell 10% of their paychecks as a kickback for him looking the other way when he signs the timesheets. If they did, Burnell should clearly reject it. The inspectors are not threatening to resign unless Burnell approves timesheets. At this point, the inspectors are not demanding anything; they are merely negotiating.

Perhaps there are some avenues available for Burnell to pay a fair amount for the services rendered by the inspectors and not become entangled with the question of whether they are reporting their hours correctly. What if Burnell crossed out the hours reported and wrote in a reasonable amount for pay in net currency? Perhaps Burnell can change the pay structure to incentivize good inspection and reporting practices, and still do so by reporting a certain number of hours worked. Clearly decisions like this may be beyond Burnell's authority, but such ideas could be proposed up the chain of command for future pay periods.

Returning to the issue of cultural relativism, we are not suggesting that because Burnell is in a foreign land he should play by their rules, even if that means compromising his values or ignoring the ASCE Code of Ethics. Canon 1 states that an engineer must "...hold paramount the safety, health, and welfare of the public..." which is not supportive of engineers' taking advantage of workers, even if the local economic climate has created a vast surplus of desperate workers. We are saying that there are cases where deeper scrutiny may reveal that what we thought were cases of corruption really are not, which is a benefit of careful, deliberate moral analysis. Burnell can then implement a solution that conforms to the ASCE Code of Ethics and compensate the construction inspectors fairly. Taking into consideration all these variables, perhaps Burnell ought to negotiate the reported hours, the hourly rate being paid, or perhaps both to a bottom line amount that both the inspectors and Burnell find appropriate and fair.

We would like to provide a dissenting opinion on this case regarding negotiating the hours reported as an illustration that some ethical dilemmas that engineers are involved with can be very complex—so much so, that even ethics experts do not fully agree on which solution is best. One of the key reasons for the recommendation for Burnell to negotiate the reported hours is based on the desire to provide a fair wage.

That reasoning is indeed noble; however, Burnell thinks about $15 per hour would be fair compensation for this work, which is more than seven times the amount the workers are actually receiving. A fair monthly wage, therefore, is unachievable, because the reported hours would far exceed the number of hours available in a week.

There are several aspects to the ASCE Code of Ethics that would argue against Burnell negotiating the reported hours: Canon 3 requires engineers to make truthful statements, Canon 4 requires engineers to serve as a faithful agent of the employer, and Canon 6 requires engineers to have zero tolerance for fraud and corruption. Burnell's approving of inflated timesheets is not being truthful, the inflated timesheets would cost the funding agencies more, and knowingly paying for hours not worked is a form of fraud.

This dissenting opinion recognizes that the local economy is completely broken, seemingly broken beyond repair. That the inspectors are being paid only up to $2 per hour and QualSpec is receiving $60 per inspection hour is ludicrous. A company receiving three times the hourly compensation of an engineer is reasonable, but this ratio is 30 times higher. It's also not Burnell's engineering responsibility to fix the unfair work environment of the Afghan economy. The workers are free to negotiate better terms with QualSpec or quit. Burnell knowingly approving inflated timesheets is a slippery slope and one that should be avoided. Burnell's implementing this hardline solution is defendable with his employer and peers, but it would cause a major disruption and conflict with the inspectors that would need to be monitored closely. It is likely that the inspectors would quit, and without inspections the construction projects would be halted. The inspectors may retaliate against Burnell by sabotaging the projects. Burnell may even be reprimanded by his supervisor and contractors for making the inspectors so miserable. This path forward for Burnell is guaranteed to be difficult and very uncertain.

Questions

1. What are the strengths of the two different business practices described (i.e., reporting work hours regardless of accomplishments and negotiating compensation based on accomplishments even when the workers are being paid by the hour)?
2. Which of the two different business practices is more grounded in ethical standards?

3. What is the justification for Burnell to implement the procedures he is familiar with in a foreign land that has very different business practices?
4. What does zero tolerance in Canon 6 mean?
5. Does the fact that Burnell is a military engineer change anything?
6. What would you advise Burnell to do? What would you advise inspectors to do?

An engineer learns of a large unexpected cost associated with shipping a custom piece of equipment across the ocean.

Case 2. The Captain's Fee

Beni Sanjay couldn't believe it. He had worked for years on a big wastewater plant improvement project, and now this ethical dilemma arose. The project was an approximately $100 million expansion of a city's wastewater treatment plant. He had ordered specialized, very large pumps from an excellent pump maker in Europe nearly a year earlier. The pumps were constructed on time and sent to a nearby shipping dock to be put onto a freightliner for the United States. That's when the problem started. Beni had received a quote from a shipping company and had noticed a

20% captain's fee on the estimate. Beni determined that the captain's fee was a euphemism for a grease payment (which is a euphemism for extortion). Beni began communicating with other shipping companies and the captain's fee suddenly jumped to 30%. Obviously, he was being forced into playing by their rules. Beni considered moving freight via land to another seaport, but he learned numerous other fees would have to be paid to get the freight out of the region. This plan was not a solution, because Beni and his firm adhere strictly to the ASCE Code of Ethics.

In addition, the city's current wastewater plant was out of compliance with EPA regulations and was being fined heavily every day that went by without correction. Downstream communities were adversely affected by the poor quality of the effluent going into the river. Beni had a serious problem on his hands. He hated for the project to be delayed, for the environment to continue to be polluted at high levels, and for the project costs to skyrocket if new pumps had to be ordered. He was stuck between a rock and a hard place.

Discussion

Beni is facing a clear case of extortion. Is this situation similar to the mail carrier demanding a fee to deliver business mail to a remote project office, or is it closer to a government official expecting a kickback to approve a large project? Clearly Beni is facing the latter: a person (or persons) in a unique position of authority demanding direct payment, above what has already been agreed on, for services that Beni is entitled to already.

What does this mean for Beni's project? Is it possible that his refusal to pay the captain's fee will result in millions of dollars of lost revenue for his company, not to mention project delays and a continued environmental impact in the community waiting for the wastewater plant project to be completed? Wouldn't paying the bribe save his company in the long run? After all, the invoice would read "captain's fee," not "grease payment," so what's the big deal? The big deal is that if Beni gives in to extortion, he does a disservice to the engineering profession. It's that simple. Not only does the captain win (and therefore continues the practice undeterred), but the entire enterprise earns a bad reputation: you can't get things done unless you give in and participate in the fraud and corruption.

In this case, Beni may have some recourse to complain and rectify the situation. Most multinational corporations have policies governing grease payments, so perhaps Beni can raise the issue with his company's leadership. In addition, a vast majority of nations in the world are signatory to the

United Nations Convention against Corruption (UNCAC), and there are methods in place to report such transgressions to the government where the shipping company operates.[7] Openly raising the flag, complaining loudly, and using the systems in place to combat corruption—including the legal system, if necessary—are all methods that Beni should consider. Of course, taking all those steps would be a pain in the neck. Yes, it may result in lost revenue for the company, but it's the right thing to do.

Questions

1. What are the ethical issues that Beni is facing?
2. What could be considered corruption in one country is legal and common in another country. Should Beni follow the "when in Rome, do as the Romans do" edict?
3. Which is worse, giving in to corrupt practices or prolonging an adverse effect on the environment?
4. What are all the options available to Beni? What are the pros and cons of each option?
5. At what point does Beni begin discussing the situation with the client?

Case 3: The Arrogant Inspector

A mega project to construct a manufacturing plant was nearly complete. The project had taken 10 years and cost $500 million. Michelle Delagardia had been on site, in a foreign country, nearly every construction day for the last 10 years. She was a lead engineer and knew every detail of her part of the plant. The day for final inspections by the government officials had finally come. Michelle's team had succumbed to the shady business practices of the region, but the entire team hated to participate in the corrupt system. The lead inspector had 20 junior inspectors and translators in his group for the final inspection. Michelle's team had given the lead inspector expensive gifts, as was customarily done in that country, upon his arrival for the final inspection, and the gifts were well received. Michelle was leading the inspection team through her part of the plant when the lead inspector began questioning a certain water treatment process. Michelle and her team explained in great detail why the process was the way it was, yet the lead inspector declared a change was needed. The intensity of the

An engineer ponders unreasonable demands of an inspector.

conversation grew into an argument with no resolution in sight. It was obvious to Michelle that the lead inspector was attempting to show his superiority and could in no way back down from what he had said, even if he was wrong. The lead inspector demanded a specific change to the water treatment process or he was not going to approve the plant.

Questions

1. What is your opinion about Michelle's company participating in unethical business practices for 10 years?
2. What are Michelle's options?
3. How does past participation in a corrupt system affect what options Michelle has?
4. What are the consequences if Michelle were to change the processes as required by this high-level governmental inspector?
5. How does the unethical conduct of the lead inspector make this an ethical dilemma for Michelle?
6. What guidance can Michelle get from the ASCE Code of Ethics?

---

[1] Olken, B. A., and Pande, R. (2012). *Corruption in developing countries*. February. <http://economics.mit.edu/files/7589>.

[2] See Søreide, T. (2014). *Drivers of corruption: A brief review*. World Bank Study. World Bank, Washington, DC. <https://openknowledge.worldbank.org/handle/10986/20457>.

[3] Kenny, C. (2006). Measuring and reducing the impact of corruption in infrastructure, Policy Research Working Paper 4099. World Bank, Washington, DC. <https://openknowledge.worldbank.org/handle/10986/9258>.

[4] U.S. federal law can be used to help enforce certain corrupt practices. For example, the Foreign Corrupt Practices Act (FCPA, under U.S.C. 15, Commerce and Trade) makes it illegal for companies and their supervisors to influence anyone with any personal payments or rewards when engaged in business transactions with foreign governments or entities. See more information at https://www.law.cornell.edu/uscode/text/15/78dd-1.

[5] Daniel Lawrence Whitney (b. 1963), see http://www.larrythecableguy.com/.

[6] Rachels, J. (1986). *The elements of moral philosophy*, Random House, New York, 15–16.

[7] For information on UNCAC, see http://www.unodc.org/unodc/en/treaties/CAC/index.html. There are other international anti-corruption and anti-bribery agreements in place to which many countries are signatories. Also, see http://www.oecd.org/cleangovbiz/for information and resources.

Chapter Eight

Continue Professional Development

Canon 7. *Engineers shall continue their professional development throughout their careers, and shall provide opportunities for the professional development of those engineers under their supervision.*

An engineer's education and advancement are never complete. This is the demanding message of Canon 7. There is no time when an engineer can sit back on his or her laurels and say, "I've arrived. I'm at my peak as an engineer and I need do nothing more."

Are the requirements of Canon 7 unreasonable? If engineers have completed their formal training and perform their work competently, why isn't that enough? Why does keeping their licenses up to date require that they must learn more? Why do they have to keep scaling new heights of engineering expertise? And if an engineer is not an engineering professor, why is he or she responsible for aiding the professional development of engineers he or she supervises? Why isn't it enough simply to supervise them?

There are two important justifications for Canon 7. The first is that engineering itself, as a practice and a body of specialized knowledge, is constantly growing. For example, an increasing emphasis on sustainability has led to many changes in design work, advancements in construction materials, and techniques to make projects more sustainable. The LEED (Leadership in Energy and Environmental Design) Certification process is a good example of the advancement made in planning communities and constructing structures of all types. Another example is the Envision rating system that ASCE has developed. Envision is focused on considering infrastructure projects in a holistic way. To remain a competent engineer, one must keep up with new technologies, practices, standards, and processes. Second, the profession itself stays healthy only if its practitioners make substantial contributions to its growth, and one of the most important contributions more senior engineers can make is to ensure that the next generation of engineers is being well-trained.

Once again, we see that becoming a professional means taking on responsibilities that go above and beyond the responsibilities of a regular employee hired into a job. A professional no longer represents only himself or herself but the profession as a whole. A professional does not punch a clock and do the minimum necessary to earn a wage, but makes a commitment to developing as a professional. Of course, hourly wage workers may commit themselves and do exceptional work—many do—but that's not always part of the work culture.

Are there any limits to how much time, energy, and commitment a professional is expected to devote to the profession? Canon 7 says only that engineers *shall* continue their professional development, but it does not say how much development is necessary. We all know engineers who seem completely identified with their work and don't take adequate time to nurture their relationships and interests outside of work. It is an easy trap, because there is always more to learn in engineering and there are always ways to improve as an engineer.

Fortunately, there are professional guidelines that communicate a consensus about how much professional development is reasonable to expect. Relicensure as a professional engineer generally requires a certain number of hours of continuing education. The requirements differ by location, but 20 to 30 hours of continuing education every two years is a common amount. This requirement is demanding but manageable for most engineers. Sometimes meeting the cost is as much of a challenge as finding the time; some companies pay for their engineers' continuing education, but some do not.

Beyond meeting relicensing requirements, each engineer must decide for himself or herself what kind of work–life balance makes sense. Psychologists who work on this issue know that a single-minded focus on work is unhealthy; however, a lackadaisical, do-the-minimum attitude toward work can also be bad for one's well-being. Between those two extremes, however, there is wide variability in what works for individuals. It would be wrong for the engineering profession to dictate a set number of hours that engineers should work.

One difficulty that arises in applying Canon 7 is figuring out which activities should count toward professional development. For example, if an engineer reads a great deal in preparation for a new project, should that count as continuing education? Should there be a cap on how many hours of reading will count? What about having a long conversation with an expert in an area that an engineer is learning about? What about continuing education on nonengineering topics, such as improving writing skills or business management skills? There is room here for a wide range of reasonable views, and ethics alone can't determine a correct answer to what should be required. Ethics can only state that engineers should be held to a high standard of development, because their responsibility to the public is so great. Ethics can also state that engineers should report their continuing education hours honestly; licensing boards set continuing education requirements, but reporting is usually done on an honor system, so engineers have to hold themselves to a standard of honesty.

Most difficulties in applying Canon 7, however, arise from its demand that engineers encourage other engineers' development. Why is this requirement even in place, and how does one go about fulfilling it?

The rationale for assigning an ethical duty to aid other engineers in their professional development is again based on the definition of a professional. A professional receives credibility, prestige, training, and support from his or her profession. In return, he or she owes it to the profession to represent it well and to provide support and training to fellow professionals. Newer members of a profession tend to receive more than they are able to give back; they benefit from the work of generations before them who have built up the profession, contributed to its growing body of knowledge, and established good relationships with the lay community. As these newer professionals establish themselves and start taking on supervisory roles, they have a duty to give more service back to the profession. Much of this work will be informal and unpaid, and it may not even receive a simple "thank you" from the people it benefits. Nonetheless, it is important work, and the profession could not thrive without it.

To aid the professional development of supervisees, a senior manager or senior engineer must focus on many aspects of the supervisees' development as engineers. A common view is that professional development includes only the development of the technical expertise of the engineers one supervises; that is, they should be given opportunity to acquire more engineering knowledge and perhaps more training in written and verbal communication and business management skills. Technical expertise and "softer" communication and business skills are indeed important, and an increasing number of engineers are pursuing graduate degrees to hone their knowledge.

However, there is much more to development as a professional than acquiring more technical and business knowledge. To be a good professional, one must be ethical, one must be effective, and one must be motivated and sustain an intrinsic love for the work itself. None of this is possible without a supportive work environment, so it should be part of Canon 7 that supervisors are responsible for creating supportive work environments so that their supervisees can fulfill their potential.

Many remain skeptical about the need for this kind of professional development, so we will give some arguments for it. First, regarding the ethical development of employees, many studies have shown that ethical behavior is not a matter of ethical knowledge alone.[1] People can memorize a professional code of ethics and even be very good at applying it to complicated situations, but having knowledge of ethics does not imply the person will be motivated to act ethically. Furthermore, these studies show that even when an individual is ethically motivated, the social situation in which one finds oneself plays at least as large a role in determining behavior as the individual's own values. People are greatly affected by the behavior of people around them and by the incentive structures in place. When appropriate behavior is regularly punished and inappropriate behavior is rewarded, even very ethical people will often start to behave badly. For example, if it is common practice to pad billable hours in a workplace, many employees will start to do that. They may not even be aware of the change in themselves. Thus, those who have the power to influence a work environment have a duty to create an ethical climate that helps people act as the profession requires.

How does one go about creating a work environment conducive to ethical behavior? An important first step is simply to make sure that people regularly discuss ethics. Discussion of ethics is often done through formal presentations, but it's important that informal discussions also take place because people need to be able to brainstorm ideas for dealing with ethically ambiguous situations that they encounter in their work. It is good for supervisors to informally highlight the ASCE Code of Ethics and encourage a brief discussion prior to starting a project. People may already be familiar with the Code of Ethics, but as we've seen in the case studies covered in this book, it's not always easy to see what the Code requires us to do in complex, real-life situations. Furthermore, people who lead ethics educational activities have discovered that even people who have worked together for a long time don't realize they have quite different ethical views and different expectations of each other.[2] These differences can lead to miscommunication or hurt feelings, even if everyone involved shares the goal of being ethical.

Another important part of creating an ethical climate is ensuring that raising concerns about ethics is always rewarded, or at least not punished. Procedures for allowing people to report concerns vary widely from company to company, but there needs to be some effective procedure in place. Unfortunately, sometimes well-intentioned practices backfire. For example, posting signs saying "X number of days without an accident!" can discourage people from reporting accidents, because they don't want to break the record. It's a good idea to conduct periodic anonymous surveys of employees to make sure whatever procedures are in place are working. Furthermore, when people acknowledge their own mistakes and try to fix them, this should also be rewarded and not punished. That doesn't mean a person should never be reprimanded for errors, omissions, or ethical or technical lapses; rather, the action of admitting and correcting the lapse should be praised, and the person should be treated with respect even if in the end it is necessary to fire him or her.

Our second argument supporting supervisors' need to focus on the general work climate and not only on developing the technical expertise of their supervisees is that being a good engineer requires more than knowledge. A person must love the field and be motivated to constantly improve, and, again, individual motivations and behavior in this area are greatly affected by one's surrounding culture. A work culture that rewards cutting corners or that treats regulations as annoyances to get around will influence everyone working there. More subtly, a work culture that is unfair, where unpleasant tasks are not fairly distributed, or some people are allowed to take credit for other people's work, or supervisors play favorites is demoralizing. Being overworked can also be a problem for people's motivation and can cause burn out, but when the high work demands are fairly distributed and people are credited for the work they do, it's much easier to avoid burnout. People will feel like they're pulling together to accomplish something difficult.

Thus, Canon 7 requires that supervisors try to create a fair, supportive work environment. Unfortunately, this objective is more difficult than it sounds. One of the major obstacles is a phenomenon that has been studied seriously only in the last 20 years or so: implicit bias. Implicit biases are unconscious positive and negative associations with various features of the world, including various categories of people. For example, one can have a positive reaction toward tall people without being conscious of it at all. This phenomenon is universal and pervasive: it affects everyone and it affects many of our behaviors. Psychologists at Harvard University have developed a measure of implicit bias called the Implicit Association Test.[3] Numerous studies using this measure have found that most North Americans have implicit biases in favor of youth, thinness, whiteness, body type, and being male.[4] These biases are widespread among people of all ages, weights, races, and genders, and they appear whether or not the person is consciously against racism, sexism, and other "isms." These biases have been shown to affect important behaviors, such as judgments about resumes, judgments about how well someone performed in a job interview, judgments about whether someone deserves a performance award, or other service or benefit. They can also affect workplace interactions, such as whether someone has the chance to speak in a meeting, whether a person's efforts are fully recognized, and whether a person is taken seriously as an engineering colleague.

The implications of this research are that creating a fair work environment is not an easy task. Even having formal workplace rules that are fair and that everyone tries to follow doesn't mean everyone will be treated fairly. Supervisors may believe they are treating all supervisees the same, but they may be unaware that they are assigning more prestigious projects to male employees. Women and minorities on work teams may not receive equal time to talk or have an equal voice in the group's decisions, which in turn gives them less experience in leadership roles, so they may not advance as quickly. Another serious problem is that when promotions or awards are being considered, a process that relies on supervisors to simply think of a list of candidates is likely to return a list that is biased in favor of white males. Studies have shown that other processes work better: for example, in a smaller company, one can simply consider every employee as a candidate. In larger groups, one can allow people to nominate themselves for the promotion or award instead of having supervisors generate their own lists of candidates.[5]

The research on implicit bias and how to counteract it is still young. Canon 7 indicates that supervisors should keep an eye on this research so they can continue to improve their workplace culture. Learning to counteract implicit bias could improve the technical competence of engineers they are supervising as well. Mental stress interferes with learning even more than physical stress, so a stressful work environment will affect employees' ability to absorb new information and make good engineering decisions.[6] A positive, fair, ethical work environment helps engineers meet their full potential.

Case 1. Boss Gets the Lion's Share of Professional Development Funds

Korry Bates, a well-established supervisory engineer, participates in ASCE committees and attends an ASCE conference each year, which is paid for by his company. He feels that he receives good value from these activities. He is exposed to new engineering topics and advanced solutions, and he also gets to network with fellow engineers. Five fairly new engineering professionals report to Korry. He almost always denies their travel requests to technical conferences, because resources to pursue professional development are highly limited, not making it possible for his supervisees to attend. To make up for this, he brings back to the office lots of information from his committee and conference experiences, and he freely shares it with his supervisees. He also encourages them to use their vacation time to go to conferences that they pay for themselves.

Discussion

At first glance, Korry seems to be violating half of Canon 7. Sure, he's looking after his own professional development, but he's not doing much for the development of his supervisees. However, there are a number of things Korry could say on his own behalf. First, he is after all a supervisor. It's especially important for him to stay up to date on the latest developments in his field so that he can provide top-quality guidance for the projects he is supervising. Second, when there are limited funds, those funds need to be spent in a maximally efficient way. Sending one senior person to a conference will benefit everyone that person supervises, and it will be much less expensive than sending all five junior people. Third, Korry is, in fact, fostering the professional development of his supervisees. He shares his knowledge with them, and he encourages them to go to conferences as well. It's not his fault that there are limited funds for travel.

Korry seems to be upholding the "letter of the law" regarding Canon 7. But is he upholding its spirit? There are a number of things that Korry seems to be overlooking when he calculates that it makes more financial sense to send himself to conferences than to send his supervisees. First, the point of conferences is not only to acquire technical information. If it were, perhaps Korry could bring all that information back to share with the group, and it would be as good as if they had gone to the conference themselves. But we could say the same thing about Korry: he could simply sit in his office and read transcripts of the different conference presentations. Why did he need to travel to the conference?

One of the primary benefits of conferences is that they bring many people with similar interests and complementary skills together. It's through networking, question-and-answer sessions, chance encounters in the lobby and elevator, and other business-related opportunities that new ideas are sparked, technical questions are cleared up, and information is imparted that attendees didn't even know they needed. Less tangible benefits are also received: excitement about new developments, a sense of purpose for being an engineer, exposure to new technical areas, a reminder of the importance of ethics, connections with specialists they may need to call on in the future. All these benefits are more important to new engineers than to more senior people. New professionals don't have a large social network yet. They haven't had as much time to think about larger issues in engineering, such as ethics or sustainability—the kinds of things that will be talked about in the keynote speeches and the social gatherings afterward. They haven't had as much practice presenting their own ideas and hearing feedback from audiences. Thus, even if Korry is right that it's especially important for him, as a supervisor, to get technical knowledge from his conference attendance, he shouldn't forget that it's especially important for his supervisees to get the less-tangible benefits that come with attending conferences.

Second, Korry seems to have overlooked the fact that he probably has more personal resources for funding his own conference travel than his supervisees do. His salary is larger, and he probably has more assets built up. His supervisees may be trying to buy their first house or paying off student loans. They may have very young children at home. Telling them to spend their own money and use their own limited vacation time for conference travel is an inconsiderate, if not intentional, thing for Korry to do!

Third, Korry seems to have overlooked some possible solutions to the problem: first, he could ask his company for more professional development funds. As a supervisor, he has the responsibility to make sure his group has the resources it needs to do its job. Has he forgotten the importance of professional development to every member of the group? Has he failed to make a strong case to his supervisor for increased travel funds? Another possibility is cost sharing. Perhaps the company could use professional development funds to cover time off for conferences while the conference attendee pays the travel costs. Of course, such funds may simply not be available, but Korry is obligated to at least try to obtain them.

Thus, to truly uphold Canon 7, Korry needs to do more to try to balance the professional development needs of his supervisees with his own development needs. Taking the lion's share of funds is not in line with the spirit of Canon 7.

Questions

1. What are the options for Korry to provide opportunities to those under his supervision?
2. Why are the engineers requesting funds to go to technical conferences?
3. Why should this be an ethical dilemma for Korry?
4. Why is it an ethical dilemma for the engineers who report to Korry?
5. What are the options for the engineers who report to Korry?

Case 2. Firm Interviewing Only Women for Engineering Jobs

Ronny Vasquez has been working for a consulting firm for about three years. There are 25 engineers working in the office, of whom three are women. The firm is currently planning to hire four new engineering graduates. Ronny agrees that the firm needs to become more diverse, but it bothers him that all the candidates being interviewed are women. It makes him suspect that being a woman was the most important aspect of a candidate's application.

Discussion

More and more companies are recognizing the value of a diverse workforce. Figuring out how to increase diversity, however, is very difficult. Laws differ greatly from country to country and between regions or states within a country. Once a manager has figured out which policies are legal, there's a further question about whether the policies will be effective or will have unintended negative consequences. A policy's effects may not be seen for a long time, so it is difficult to tell if it is working. In the meantime, individual employees like Ronny will wonder if his company is acting in a fully ethical manner.

The ethics of increasing diversity cannot be figured out without understanding why diversity is valuable. There are both ethical and pragmatic reasons to be in favor of diversity within professions and individual workplaces. First, speaking ethically, fair treatment of all people is a fundamental, ethical value. Everyone deserves the opportunity to fully develop his or her talents and pursue his or her chosen career. When a profession or workplace does not reflect the demographics of the larger population from which it is drawing, it raises an ethical red flag. Sometimes the red flag is easy to lower again. Why is the average height of NBA players so different from the average height of the general male population? Because height is very important to being able to play basketball well, but height is distributed mostly by nature and there is very little a person can do about it. There's nothing immoral about selecting for height in choosing players, even though that means many men will never be able to play for the NBA, no matter how much they might want to.

Similarly, there is nothing wrong with the engineering profession selecting for engineering skills, which involve technical expertise, critical thinking skills, mathematical ability, advanced education, and other rigorous qualifications. But why are the demographics of the engineering profession so different from the demographics of the general population? Focusing on the United States, the U.S. Census Bureau says that in 2011, 16.3% of people who listed their first degree as an engineering degree were women and 4.5% were black. Those numbers are far away from the general workforce numbers; nearly 50% of the U.S. workforce is female, and 11% of the workforce is black.[7] To see that disparity as unproblematic in the same way the NBA disparity is unproblematic, we would have to think that engineering ability or drive is simply naturally distributed more toward white males than toward black males or women of any race. There's ample evidence to the contrary.[8]

The demographics of engineering firms don't mean the engineering profession is unfairly discriminating. It's possible, and likely, that much of the demographic disparity is caused by inequalities and biases within the larger society. Black people in the United States, as a group, have less access to educational resources. Girls are often discouraged from a young age from pursuing mathematical or technical interests. None of these problems is the fault of the engineering profession, but the profession should be careful that it isn't contributing to the racial and gender disparities by reinforcing stereotypes of what a "real" engineer looks like or by targeting recruitment efforts only at prestigious schools that are less likely to have diverse student bodies. Researchers studying gender disparities in engineering have pointed out that environmental and civil engineering attract more women than mechanical and computer engineering; some speculate the reason for women's greater attraction to civil and environmental engineering is that women tend to care more about helping people than about building things.[9,10] If this is true, then engineering as a profession should emphasize to young people the ways engineers help people.

Still, even if much of the disparity is caused by larger social forces, some of it may be due to discrimination within the workplace. Any company that lacks diversity needs its executives to ask themselves if their company really is doing all it can to give equal opportunity to every qualified person who might apply for a job there, who does apply for a job there, or who accepts a job there. In previous eras, recruitment was often done by word-of-mouth, which meant engineers hired people who were already in their social circle. People outside that circle wouldn't even hear of the job. Now companies are more careful to advertise in a wide variety of venues. United States law also allows "targeted recruitment," directed at underrepresented groups, which may be how Ronny's firm ended up with a short list of all female candidates.

Once candidates apply, are they treated fairly? How about after they are hired? These are two times when implicit bias can create problems. Companies should be sure they are following best-practices recommendations from agencies such as the U.S. Equal Employment Opportunity Commission (EEOC).[11] A company culture where women or minorities are ignored, mentored insufficiently, or stuck with an unequal amount of unpleasant work can make it very difficult to maintain diversity, even if the company hires a diverse workforce.

We have seen ethical reasons to increase diversity and to worry about the lack of diversity. There are also pragmatic reasons to follow this advice. First, studies have shown that diverse teams tend to be more creative and efficient.[12] Second, retention of minority and female employees improves if they don't feel isolated, as if they are token minorities, and improved retention reduces hiring and training costs. Third, if engineering potential is distributed evenly across racial groups and genders, then by not diversifying the workforce, the profession itself is missing out on a great deal of talent by not encouraging and cultivating that talent wherever it exists.

To return to our case study, Ronny already accepts the reasons to be in favor of increased diversity within his profession and company. He is more worried about the methods of achieving it. If increasing diversity becomes a top priority, it could overshadow the more basic need to hire the most-qualified people. Compare the NBA case: a head coach who focused only on height might not sign the best players. Conversely, it could turn out in any given year that the best players are also the tallest ones. If so, we wouldn't be able to tell from his picks whether he was focusing on overall skill or just height. Similarly, Ronny doesn't really know why his firm ended up with an all-female short list. Maybe in this batch of applications, the top people just happened to be women.

Ronny is certainly entitled, however, to worry about whether his firm is hiring the best people it can. He should also be concerned about whether the firm is being fully ethical and legal in its hiring practices. United States law forbids discrimination in hiring, which means companies may not discriminate based on gender, race, and several other protected categories.[13] It is possible Ronny's firm violated this law. Conversely, companies that are governed by the EEOC must advertise widely, but they can also target recruitment efforts at underrepresented groups.* Furthermore, they can target educational opportunities such as scholarships and internships for underrepresented groups. Many companies have such programs at colleges and universities, and these programs can lead to students applying later for jobs with the company, which is a legal way Ronny's firm might have ended up with so many female applicants.

Because Ronny is concerned about the situation, he should schedule a meeting with his supervisor and perhaps with the human resources specialist to find out about the firm's hiring procedures. If he discovers a potential violation of the law, he can discuss it with his supervisor or report it to the EEOC. Every company covered by the EEOC (coverage depends on the type and size of company) must have a notice about employment laws and the EEOC's contact information posted in a conspicuous location.

The worst option Ronny could take is to stay quiet but retain his suspicions. His skeptical attitude toward the company's new hires would influence his behavior toward them, even if he tried to hide it. Many minorities and women complain that part of the chilly climate they experience in their schools and workplaces comes from the opinion of others that they aren't qualified and are employed only because of affirmative action. This feeling is demoralizing and makes retention of minority students and employees difficult.

---

* Not all employers in the United States are covered. Consult the EEOC website at https://www.eeoc.gov/employers/coverage.cfm.

Questions

1. Is it important to have a diverse work force?
2. Is it important to have fair employment practices? How do you define fair?
3. How can Ronny voice his concerns about what seems to be an unfair application process?
4. What are the challenges with placing high importance on increasing diversity?
5. What are the challenges with hiring the most qualified, disregarding diversity interests?
6. Should inclusion and diversity topics be included as a canon in an engineering code of ethics?

A design engineer working late into the night to satisfy the demands of office manager.

Case 3. Overworked and Stressed Out

Lucy Neill started work a few years ago at her first engineering job. She is working for a local engineering firm where the workload always seems to be feast or famine. In the beginning, she was fine with working long hours for a few weeks at a time to complete a proposal or to meet a rapidly approaching deadline. She understood that it can be hard to predict when jobs will come in. However, she is now in the situation of having spent six straight months with a harvest-hours-type of workload: working weekends, working late into the night on a regular basis, and being called in at the last minute. Lucy's supervisor, Phil Canon, keeps reassuring her that the workload will last only a bit longer and then the firm will be able to handle a slowdown because they have been so successful in getting jobs lately, but he has said the same thing for several months. Lucy likes the work and the people, but she's starting to wonder if the company is taking advantage of her.

Questions

1. When does too much work become "too much" work?
2. Is the way Lucy's firm works simply how consulting firm environments are, or is Lucy being taken advantage of?
3. Does the situation change if Lucy received additional compensation for hours worked over 40 hours? What if she is not receiving additional compensation?
4. Is Phil being honest with Lucy?
5. Is there an increase in the potential for errors and omissions in their design work given the extensive work schedule?

---

[1] Hoyk, R., and Hersey, P. (2008). *The ethical executive: Becoming aware of the root causes of unethical behavior*, Stanford University Press, Stanford, CA; and Bazerman, M. H., and Tenbrunsel, A. E. (2012). *Blind spots: Why we fail to do what's right and what to do about it*, Princeton University Press, Princeton, NJ.

[2] Whitbeck, C. (2004). "Trust and the future of research." *Phys. Today* 57(11), 48–53.

[3] Greenwald, A. G., McGhee, D. E., and Schwartz, J. L. K. (1998). "Measuring individual differences in implicit cognition: The implicit association test." *J. Pers. Soc. Psychol.*, 74(6), 1464–1480.

[4] Banaji, M. R., and Greenwald, A. G. (2013). *Blindspot: Hidden biases of good people*, Delacorte, New York.

[5] Lincoln, A. E., Pincus, S., Koster, J. B., and Leboy, P. S. (2012). "The Matilda effect in science: Awards and prizes in the US, 1990s and 2000s." *Soc. Stud. Sci.*, 42, 307.

[6] Avey, J. B., Luthans, F., Hannah, S. T., Sweetman, D., and Peterson, C. (2012). "Impact of employees' character strengths of wisdom on stress and creative performance." *Hum. Resour. Manage. J.*, 22(2), 165–181.

7 Landivar, L. C. (2013). "Disparities in STEM employment by sex, race, and Hispanic origin." American Community Survey Reports, *ACS-24*, U.S. Census Bureau, Washington, DC.

8 Hyde, J. S., Lindberg, S. M., Linn, M. C., Ellis, A. B., and Williams, C. C. (2008). "Gender similarities characterize math performance." *Science*, 321, 494–495.

9 Myers, J. (2010). "Why more women aren't becoming engineers." *Globe and Mail*, Nov. 9.

10 Jacobs, J. E. (2005). "Twenty-five years of research on gender and ethnic differences in math and science career choices: What have we learned?" *New. Dir. Child. Adolesct. Dev.*, 110, 854.

11 U. S. Equal Employment Opportunity Commission (USEEOC). (2015). "Best practices of private sector employers." <http://www.eeoc.gov/eeoc/task_reports/best_practices.cfm> (Nov. 17, 2015).

12 Phillips, K. W. (2014). "How diversity makes us smarter." *Sci. Am.*, 311, 4.

13 U.S. Equal Employment Opportunity Commission (USEEOC). (2015). "Best practices of private sector employers." <http://www.eeoc.gov/eeoc/task_reports/best_practices.cfm> (Nov. 17, 2015).

Chapter Nine

Complex Case Studies

This book has presented a number of case studies that each bring out the complications in a single canon. In real life, however, we often run into situations that involve many different kinds of ethical concerns. Some of these concerns may conflict with each other. Therefore, we would like to leave the reader with two cases that can be addressed only by drawing from multiple canons. Both cases also highlight the sustainability clause of Canon 1, because the directive to "comply with the principles of sustainable development" is especially difficult to know how to apply, making these cases particularly challenging.

Complex Case 1. Unintended Consequences (or Murphy's Law Strikes Again)

Kyla Brookes, a civil engineer, works at a consulting firm that focuses on land development in the Murphy River watershed. For years, there have been lawsuits against the city of Murphyville alleging that development is causing an increase in streamflow downstream. The streams in the watershed were once full of aquatic life. There were fish in the pools and a healthy shallow shoreline for a thriving plant and animal environment. The stream would reach bank capacity after a heavy storm only about once a year. The downstream property owners relished the diverse and healthy stream environments on their land. They had trails and picnic areas along the shoreline, and they thoroughly enjoyed living by this stream. In a way, the healthy stream environment was the most valued aspect of the properties.

However, as the watershed became increasingly developed, the downstream landowners started complaining to the city officials that their once mild mannered stream was turning into a raging river much more frequently because of the developments that had been constructed upstream. The landowners also complained that the stream was changing. The shallow shoreline areas were being eroded away, stream crossings were being damaged because of the higher flows, and a rectangular stream channel with vertical banks was all that remained.

All of the approximately 20 residential and commercial developments in the watershed were developed according to local regulations. Unfortunately, storm water flow for the watershed as a whole was never considered. Approximately 20 detention basins were built in the watershed, with each basin being designed for each specific development. Kyla designed four of the detention basins.

Spurred on by the lawsuits, the city officials ultimately voted to fund a watershed modeling project for the entire watershed to determine what the effects of the individual detention basins were. It was determined that downstream flood flow rates from all the developments were actually higher than they would be if no detention basins had been built. It was also determined that streamflow during more-frequent rainfall events was much higher than in predevelopment conditions. These changes were caused by the detention basins changing the timing of the flood flows. Plus, the watershed now had far more hard surfaces that prevented infiltration and increased runoff. Frequent rainfall events were not controlled, because detention basins were primarily designed to reduce flood flows, and the developments desired virtually no standing water in the basins. Because the low flows were practically bypassing any detention, the downstream flow for the more-frequent rainfall events was greatly increased.

Kyla has stayed informed on this situation. The downstream residents have been featured numerous times in the local newspapers. Some of the residents said they were sick over what had happened to their stream, to their property, to their way of life. Kyla knows the results of the modeling project and also knows that the city officials have quietly settled lawsuits to reduce litigation risk.

Kyla is now determining stormwater solutions for another development in the same watershed. Her firm and the city officials want the same solution as before (i.e., design an individual detention basin only for the local development without consideration of the entire watershed). Kyla wants to design a detention basin that rightly considers the entire watershed, but when she proposed this design to the developer, the developer said "no way" because of the added expense.

An engineer considers financial, social, and sustainable topics associated with available design solutions.

Discussion

If you were Kyla, facing such a complex situation, would the ASCE Code of Ethics be the first place you would look for help? Perhaps not. The code contains individual canons that each seem to fit with paradigm situations: a dangerous bridge, a bribe being offered, and the other cases covered in previous chapters. None of the canons seems to address a situation where there are multiple, competing interests and considerations. But as we've seen throughout this book, each canon contains complex concepts that can be used to help sort through difficult issues. If we bring the relevant canons and their concepts to bear on Kyla's situation, we may gain some insight.

Which canons seem to apply here? First, and most obviously, Kyla knows her actions will have a long-term effect on the public and on the natural environment. Any one detention basin may have little impact, but each one contributes to the larger problem. Thus Canon 1 seems very relevant. Canon 2 regarding working in one's area of competence also applies, although the connection is less obvious. Kyla is a civil engineer, qualified to design solutions for streamflow problems. However, she is neither an environmental scientist nor an ecologist and may know very little about the impact of different water flow patterns on native species of plants and animals in the region. Furthermore, if she has mostly focused on individual developments and their needs to control water, she may have little experience with designing holistic solutions for entire watersheds. Is she really competent to speak as an engineer about what the community as a whole needs? As Kyla decides whether to speak up about the issue, she also needs to keep Canon 3, on public communication, in mind. Finally, Canon 4, regarding being a faithful trustee, is very relevant. Kyla's own firm, the developer, and the city officials have their own settled opinions about what ought to be done; Kyla needs to be sure that if she tries to convince them to make a different decision, she does it in a way that is faithful to her role as an employee of her firm.

When more than one canon is in play, how can we deal with potential conflicts between them? Which canon should be prioritized over which? For example, Kyla is in a situation where public well-being and sustainability are at stake, so Canon 1 would seem to tell her to do as much as she can to improve the situation. But some steps she could take to improve the situation might violate other canons. Perhaps she would need to step out of her area of competence to help the public, or maybe she would have to act against her own firm's interests to help the public. Does the Code of Ethics help sort out conflicts like this?

The ASCE Code of Ethics does not contain a ranking of canons, except that Canon 1 does state that public welfare shall be held "paramount," which indicates that public welfare is especially important. However, Canon 4 states that engineers shall act as faithful trustees, which is a deontological principle that does not allow exceptions, so faithfulness seems equally important.

Can stepping back to ethical theory help us out here? Yes, to some extent. Philosophers who have developed ethical theories have been deeply aware of the potential for moral dilemmas and conflicting rules, and they have tried to build solutions to those problems into their theories. *Consequentialism* has the most straightforward way of dealing with the problem: every state of affairs contains a certain amount of happiness or welfare within it, so states of affairs can be ranked according to how much happiness they contain. Consequentialism demands that one try to produce the best state of affairs, taking into account long-term consequences, so in principle there should be no conflicts. If two states of affairs are equally the best, then it's permissible to work toward either one of them. In Kyla's case, there should be some answer to the question of whether it would produce better results overall for her to violate one canon for the sake of upholding another canon. If it would, then that is the morally correct thing to do according to consequentialism.

Kantian ethics and *virtue ethics* deal with potential conflicts in a less quantitative way. In Kantian ethics, failing to respect someone's rationality is always forbidden: one may never lie, defraud, or break contracts with others. Promoting other people's good, however, is not always required. If the only way to promote welfare is to violate someone else's autonomy, then Kantian ethics requires the agent not to promote welfare in that case. In conflicts between Canon 1 and deontological canons, such as Canon 4, Kantian ethics would prioritize the deontological canon over Canon 1.

Virtue ethics is even less quantitative. There is no ranking at all the different considerations a virtuous person will take into account in his or her deliberations. Rather, "practical wisdom" is supposed to help a virtuous person find creative solutions that do justice to all the different morally relevant considerations. Practical wisdom, however, is not easily acquired. A truly wise person would possess all the virtues (such as justice, benevolence, and courage, among others) so that he or she could be sensitive to all relevant moral concerns. But no actual human being possesses all the virtues, so it is quite difficult to know how to apply virtue ethics.

Adding to the complications here, there is no grand meta-theory that tells us which of these specific ethical theories takes precedence when they give conflicting answers to a moral problem. Even if consequentialism and Kantianism give definitive answers to the question of what Kyla should do in this situation, those answers might conflict which does not help Kyla's situation.

Trying to use ethical theories directly, then, doesn't seem to help. However, as we discussed in the introduction in Chapter One, we can take a pluralist approach to ethical theories. Each theory is correct in its own context, and each correctly emphasizes important moral concerns. The pluralistic approach gives us the most helpful way to use ethical theories, which are very good for bringing out morally relevant considerations in various types of cases. In fact, this entire book uses ethical theories to bring out the morally relevant features of each of ASCE's ethical canons. All cases that involve public welfare, for example, must confront the thorny problem of how to weigh competing public values.

All cases that involve public communication must deal with the difficulty of drawing lines between honesty and dishonesty. Kantian ethics has brought out concerns about autonomy in applying each canon, consequentialism has brought out concerns about valuing outcomes when applying each canon, and virtue ethics has brought out concerns about acting wisely when applying each canon to difficult cases.

A useful approach, then, would be to take the canons that apply to Kyla's case and think about the morally relevant considerations that typically arise when applying those canons. Thus, to analyze Kyla's case, we go back to the discussions of Canons 1, 2, 3, and 4. Once all the relevant considerations are on the table, it will be easier to see how to put them together into a responsible decision.

Returning to the discussion of Canon 1, recall that one justification for making public welfare paramount is that imposing risk on people requires their consent to the risk imposed. Because individual consent is impossible to get for all our actions, we can use the law itself as a proxy for consent, because in a democratic society the law reflects the will of the people. The law is also important for weighing competing values; because reasonable people can disagree about how to prioritize different values, democratic decision making tries to allow everyone to be heard. Besides the law itself, government agencies, such as the Environmental Protection Agency (EPA), can serve an important role in reflecting public opinion (see Canon 1, Case 2).

Applying all of this to Kyla's case, Murphyville's regulations regarding detention basins and land development indirectly reflect the will of the voting population of Murphyville. There is no evidence in this case that Murphyville is corrupt or not following a democratic decision-making process, so all citizens have, in principle, had a chance to be heard and to voice their own preferences regarding how to prioritize issues such as the expense of different kinds of development versus the beauty of the river's banks. Kyla has a very good reason to take the existing regulations of Murphyville seriously.

However, as we discussed in the chapter on Canon 1, the democratic process is not perfect. People do not reason perfectly, even about their own welfare; it is possible for the public to make bad decisions and end up harming themselves. Also, government agencies are not perfect in the way they carry out their jobs. If Kyla has good reason to think the city officials have made bad decisions in formulating the regulations, or that those officials didn't take their own modeling project seriously enough, she is not obligated to stand by and defer to "the will of the people."

Her design project is going to impose risk on the public, and she needs consent before doing that; if she has reason to doubt that good, fully informed consent has been given, she has the moral right, and even the moral responsibility, to speak up. Furthermore, consulting the appropriate state agency, Federal Emergency Management Agency (FEMA), or the EPA would not be a violation of her duty to respect the citizens of Murphyville. As discussed in Chapter One, these government agencies also reflect the will of the people in general to protect their natural environment, and they have resources and expertise to make judgments about larger environmental issues. The EPA, for example, may well be in a better position than the local government to judge what's best for the watershed as a whole.

Kyla has good reason to investigate further and speak up. However, issues raised by Canon 2 will tell her to be cautious in her approach because the canon requires that she perform services only in her areas of competence. If she speaks as an authority on environmental issues without having the relevant training, she can undermine trust in the engineering profession and do more harm than good. As we know from the discussion of Canon 2, it can be difficult to tell when an engineer is performing a service. Engineers have free speech rights like everyone else, but they must be careful to indicate when they are giving an official engineering judgment and when they are speaking simply as concerned citizens.

It may be impossible, however, for Kyla to speak purely as a concerned citizen. She is, after all, the original designer of some of the current detention basins, and she is working for one of the companies designing future detention basins. Because of her engineering experience and expertise, she has a much more nuanced view of why the existing regulations are inadequate than the average citizen does. Furthermore, she doesn't want her voice to carry the weight of merely a concerned citizen; if the landowners being most affected by the development have not been able to convince the city officials to change the regulations, a random citizen who doesn't even own property in the affected area can't expect her opinion to carry much weight. Kyla needs to draw on her status as a civil engineer if she wants to effect real change.

But if Kyla is going to speak up as an engineer to change the existing regulations, she must be sure to speak only in her area of competence as Canon 2 requires. Furthermore, Canon 3 will step in to tell her she must also be truthful and objective in her statements. As we saw in the discussion of Canon 3, being truthful and objective would require that

Kyla be aware of her own biases in this situation and try to keep them from coloring her public statement. Her goal should be to deliver information in a way that allows the citizens of Murphyville to reason well by using their own wisdom about the situation, rather than necessarily reasoning their way to Kyla's preferred conclusion. At the same time, she needs to consider the competence of the people with whom she speaks; they may not be able to understand the data in the same way that she does, and she needs to make the data accessible to them. When she is communicating with the public about the data, she might be tempted to color the data to persuade people to reach her preferred conclusion, so she needs to be especially careful how she presents the information she wants to give to the city officials and the public. Furthermore, Kyla is in a similar situation to Dr. Gonzalez's in Canon 3, Case 2 (Chapter Four): she has access to specialized information that the public does not have, and she has expertise in civil engineering. Thus, she can't help speaking as an engineer; as an authority on these issues, her voice will carry extra weight even if she tries to assure people she is only speaking as a concerned citizen. She needs to be especially careful to avoid bias in the way she presents information.

Finally, Canon 4 tells Kyla to be faithful to the interests of her employer and clients and to avoid conflicts of interest. As we know from our discussion of Canon 4, it can be difficult to determine the true interests of one's employer and clients. The interests may not be stated explicitly or even known by the person for whom one is acting as a trustee. Kyla's employer and client certainly want to control costs and stay within the law, which would explain their desire to follow the existing city regulations. However, if future lawsuits are not able to be settled quietly, the city officials could easily turn on the developer and try to place blame for harmful effects of development. The developer could then shift the blame to Kyla's firm for not warning the firm of the risks associated with the detention basins. An engineering firm is supposed to use its expertise and provide guidance for its clients, not simply do as it's told. Furthermore, a firm can benefit from gaining a reputation for innovation and forward thinking. If the firm took the lead in environmentally friendly design, the initial investment could pay off handsomely in the long run.

Kyla's client is the developer, but the developer has an interest in keeping the city officials happy, and we have already discussed the possibility that the officials are not making good decisions about Murphyville's future. It may well be putting too much emphasis on short-term costs over long-term sustainability. Because Kyla's firm, the developer, and the city officials could all benefit from a more environmentally sophisticated approach to development, it is not necessarily disloyal or unfaithful for Kyla to speak up and try to show them these benefits. She would still be acting for the good of her employer and client.

Is Kyla in a conflict of interest situation? If her interest in sustainability is not truly in conflict with her duty to be faithful to her employer and client, then it looks as though she isn't. However, recall Georgio in Canon 4, Case 2 (Chapter Five), who experienced a conflict between his need to exercise objective engineering judgment and his desire to see the best products developed. Such a conflict can cause bias to creep into an engineer's judgments. Kyla is in a similar situation. As she considers what to say to the city officials, to her employer, and possibly to the EPA, she will be tempted to emphasize data that support her desired outcome and deemphasize data that conflict with what she wants. She needs to stay vigilant about this possible bias, and she should inform those with whom she is speaking that she is motivated by concern for the environmental effects of the existing regulations, because this motivation may be coloring her judgments.

We have now discussed the morally relevant considerations raised by Canons 1 through 4 for Kyla's situation. An interesting question, however, is whether the ASCE Code of Ethics is complete. Are there morally important factors that are not captured by any of the canons? This case brings out two such factors, raising the issue of whether the code ought to be expanded or refined.

First, much of Kyla's dilemma arises from the problem of unintended consequences. *Unintended consequences* happen when one acts in a way that makes sense when looking at a situation in isolation, but that ends up making things worse when one's actions are combined with other people's actions. (These are also sometimes called *coordination problems*.) For example, walking across a lawn does little damage to the grass, so there's no reason not to take a shortcut across a public lawn. But if many people take the same shortcut, a path of dead grass will be carved in the formerly beautiful lawn. The only way to prevent this is to impose on everyone an incentive not to walk across the lawn, even though any one person's walk won't harm the grass and allowing a small number of people to walk across the lawn wouldn't be a problem. This rule regarding keeping off the grass can seem unfair to people.

The city of Murphyville does not seem to take the problem of unintended consequences very seriously. Their regulations are based on the idea that if each developer does what is best for his or her development, the combined actions of all developers won't be a problem. Perhaps they're worried about putting unfair burdens on each developer. After all, it would be expensive for each developer to consider the watershed as a whole, and yet no one development is causing serious problems. Why should they be forced to pay to fix a problem they didn't cause?

Should engineers be held responsible for looking at the larger picture and thinking about how individual actions can combine to produce surprising results? If anybody would be good at analyzing complex situations to see how the parts create a whole, engineers would. But nothing in the ASCE Code of Ethics requires engineers to consider the larger picture in this way. They are tasked with considering public welfare and sustainability, but these concepts are vague. Perhaps the code should include explicit reference to considering unintended consequences and watching out for coordination problems. These concerns could be added to the list of subprinciples under Canon 1.

Second, some of the effects on public welfare that Kyla is concerned about are qualitative and difficult to measure. The landowners are not only concerned about species diversity or ecosystem health; they miss their beautiful environment. They want the burbling stream and chirping birds back. These are aesthetic concerns, concerns with the appearance and feel of one's environment. Should engineers try to preserve and promote beauty in their projects in addition to making projects functional and efficient?

It might sound strange, but there are professions that include aesthetic concerns in their codes of ethics. The code of ethics of the American Institute of Architects includes the statements, "Members should continually seek to raise the standards of aesthetic excellence, architectural education, research, training, and practice," and, "Members should respect and help conserve their natural and cultural heritage while striving to improve the environment and the quality of life within it."[1] The American Society of Landscape Architects has similar statements within its code of ethics, requiring upholding both aesthetic values and the value of cultural heritage.

If ASCE's Code of Ethics contained similar requirements, Kyla could refer directly to them to make the case that development in the Murphyville watershed demands better engineering solutions for water management. The aesthetic concerns of the downstream homeowners would call on her engineering abilities directly. In other engineering projects, cultural heritage could also play a role in guiding the engineer's judgments. In certain projects, cultural heritage or aesthetics are a requirement.

Would it be a good idea for engineers to include aesthetic and cultural requirements in their codes of ethics? Doing so would mean redefining the purpose of engineering, and the profession would have a different self-image. Engineers themselves need to define what their profession means.

Drawing these different considerations together, what should Kyla do? First, she should make sure she has the competence to speak to the larger environmental impacts of her current engineering work before she speaks. She has already done some research, but she should make sure she understands why traditional detention basins are having the effects they have. A good place to start would be with the city's modeling project, which should be obtainable through a Freedom of Information Act request if it's not already easily accessible to the public. Kyla could contact the people who did the work on the modeling project to talk to them about how they reached their conclusions.

Once she's sure she is competent to speak as an engineer about these issues, she should notify her employer that in her engineering judgment, detention basins are an inadequate long-term solution to water management problems in these developments. Her supervisor or her firm's representatives will likely say that it's not their problem; they are following the city's regulations. Kyla, however, is more than an employee. She is a professional and should exercise her own judgment. She needs to tell her employer that she wants to engineer a solution that works better than the detention basins. Her employer may threaten to remove her from the project or even fire her. Should Kyla give in and fall on that sword? Not necessarily. There isn't imminent threat of harm to the public if she completes the current project as her employer wants.

However, Kyla has a larger ethical duty to try to convince the city officials to change the regulations so that the public and the environment are better served. Canon 1 requires it. So, after informing her employer of her motivations (so the employer understands the potential conflict of interest she faces), she should contact the appropriate state agency and tell them her considered judgment about the detention basins. Now that she has the competence to speak as an engineer on this issue, she can also write a letter to the editor of the regional newspaper and speak at city council meetings. If she is not successful in convincing the city officials to change, her personal ethics may prompt her to stop doing her current work and use her engineering skills on different kinds of projects or in a different region altogether. The ASCE canons, however, don't imply that she must do that. They only require her to try to improve the situation. As a member of the profession, Kyla may also want to speak up to ASCE about the Code of Ethics itself; if it does not include everything she needed to draw on in this difficult situation, she may want to press for more ethical guidance on coping with unintended consequences and the aesthetic effects of one's engineering work.

Questions

1. Why is this situation bothering Kyla?
2. Should Kyla proceed with a typical design that will be approved by the city officials? Why or why not?
3. How does the consideration of sustainable design affect this situation?
4. What are the obligations of a city engineer to rectify this situation?
5. How do the adverse effects on downstream residents affect Kyla's responsibilities?
6. Should engineers add aesthetic or cultural heritage values to their ethical guidelines? Why or why not?

Complex Case 2. The Unforeseen

The Unforeseen is a movie about the dilemma generated by two interests: a desire for urban development on the one hand and the need to protect the environment on the other.[2] The particular case detailed in *The Unforeseen* is the development of a neighborhood in Austin, Texas, called Barton Creek, which threatened the quality of a massive natural spring—and popular Austin hangout—called Barton Springs. This natural water feature was considered a sacred site by the Tonkawa Native Americans long before European settlers came to the area.[3] Settlers recognized the value of this natural wonder in the 1800s, and a private owner gave the springs property to the city of Austin in 1918.[4] About 800,000 visitors every year utilize Barton Springs for recreational enjoyment. The movie details the legal battles fought by the developers, environmentalists, and the public between 1990 and 1996, and tries to explain why in this case the private interest in the development of Barton Creek ended up hurting the public interest in environmental stewardship.

Visitors enjoying Barton Springs on a hot summer day (David Ingram 2013 No alterations made. CC BY-NC 2.0. https://www.flickr.com/photos/dingatx/9270762048/. *This image on flickr is free to share and adapt*)

We encourage readers to watch the movie. It is well done and has the potential to stimulate significant discussions around the ethical dilemmas involved. We are providing some key thoughts and questions that can be used to discuss this case study.

The main characters in the movie are as follows:

- Gary Bradley, developer
- Bill Bunch, environmental lawyer
- Dick Brown, lobbyist for Freeport-MacMoRan, Inc., a natural resources company
- Robert Redford, actor and environmentalist
- Marshall Kuykendall, real estate broker and president of Take Back Texas, Inc., an organization dedicated to pushing for property rights

Other characters include Ann Richards (then governor of Texas), Earth First (environmental group), Henry Brooks (rancher), Curtis Peterson (farmer), William Greider (journalist), and even Willie Nelson (country singer).

Certainly, several engineers were involved with this development. The engineers, however, are not highlighted in the movie.

Moral Aspect

Clearly the main moral dilemma examined by the film is the desire for urban development (pitched as a private, for-profit interest) versus the imperative to protect the environment (couched as a public interest). In a nutshell, Gary Bradley heads the story in favor of the former, accompanied by Dick Brown and Marshall Kuykendall. Arguing the environmental concerns are Bill Bunch, Robert Redford (introduced as an environmental activist rather than an actor), and William Greider, along with various accounts told by a farmer, a rancher, Gov. Ann Richards, and others. At the end, the argument is made that "with proper accounting" both development and environmental protection are achievable. Today we could call this *sustainable development*. The question is how do we achieve such proper accounting. By better—or more—government regulation?

But there are other moral aspects of this story. One that comes to mind is the question about whether "grandfathering" is ethical. Dick Brown lobbied for legislation that would have allowed developers to perform their work under less- restrictive environmental regulations, because, as the argument goes, it is unfair for the government to "change the rules in the middle of the game." Governor Richards, however, vetoed the bill. The developers, then, changed the strategy and couched the issue as a "property right" battle, a move that mobilized Texas' ranchers and farmers against the environmentalists. In 1995, soon after George W. Bush became the new governor of Texas, House Bill 1704 was passed into law, scoring a victory for the developers.

Questions

1. It does seem unethical to change environmental regulations during the time developers were trying to put together their plans (how will the developers know what to plan for if the rules keep changing?), but does this mean that we can ignore potential environmental harms that result from lax environmental regulations?
2. Another major moral issue is raised by the group Take Texas Back regarding private property. To what extent can the government dictate what individuals can and cannot do with the property they own? Is property ownership tantamount to a carte blanche on what can be done or is there a sense in which the government has a vested interest in regulating what happens in the environment, regardless of who owns the land?

3. The engineers who were involved with the design and construction of the streets, sewers, water distribution, and other infrastructure certainly knew of the local resource of Barton Springs and how undesirable it was to negatively impact the Springs' water quality. What were the engineers' responsibilities to protect Barton Springs? Were the engineers morally obligated to implement advanced erosion control techniques and to decrease the population density beyond required environmental regulations if they knew the governing regulations were not enough?
4. What can engineers do to hold paramount the safety, health, and welfare of the public when profit of development is greatly jeopardized by needed design aspects?
5. Which ASCE canons help the engineers involved in this project to resolve potential moral dilemmas?
6. As raised by Marshall Kuykendall, what happens when regulations result in a significant reduction of profit for the real estate broker? He explained that he used to be able to break down big plots into smaller ones, but that regulations required him to sell bigger lots, which resulted in a loss of $5 million in profits. Who is responsible for that? Or is it just the gamble inherent in the business of real estate brokerage?
7. Lastly, the film goes into some detail about the economic conditions under which Bradley had to operate. At the beginning, development seemed to be headed toward significant profits and success, but the savings and loan crisis of the time eventually crippled much of the banking industry and we could argue that this was clearly beyond Bradley's control. Bradley declared bankruptcy (he owed $73 million to investors and to the government) and lost the court battle. Was Bradley a scapegoat? Despite his self-admission that he did not exercise proper accounting, should Bradley carry the entire blame or responsibility?

---

[1] The American Institute of Architects (2012). *2012 Code of ethics and professional conduct.* <http://www.aia.org/aiaucmp/groups/aia/documents/pdf/aiap074122.pdf> (Nov. 6, 2016).

[2] <www.theunforeseenfilm.com/trailer.htm>.

[3] <https://en.wikipedia.org/wiki/Barton_Springs>.

[4] <https://en.wikipedia.org/wiki/Barton_Springs_Pool>.

Chapter Ten

Conclusions and Resources

We hope you have found these case studies interesting, valuable, and thought-provoking. The engineers in these situations faced difficult and challenging environments in which to perform their engineering duties.

We have a few additional suggestions for you to consider when facing ethical dilemmas. None of these always ensures complete adherence to an engineer's code of ethics, but they may stimulate some thoughts that may be helpful. They are as follows:

- Review the appropriate engineer's code of ethics periodically, maybe at the beginning of each new project or annually.
- Use the ASCE Code of Ethics or other codes of ethics as ammunition when arguing in favor of an ethical solution to a dilemma.
- Is the decision you are considering defendable? What if the local newspaper stated your decision as the headline? It might be controversial, which is fine, but is it defendable?
- What would a person you admire think of your decision or course of action?
- Consult with a trusted individual outside of your specific work environment to get a second opinion. Much insight can be gained when we describe an ethical dilemma to others, and you may receive wise counsel on the situation.
- Use one of the ethics hotlines that are available (ASCE Ethics Hotline: 800-548-2723 x 6159).
- Pursue the development of an ethical solution like an engineer. You have extensive analytical skills; don't be persuaded to quit thinking like an engineer when someone is pressing you to accept a solution that feels like it violates your engineering code of ethics.
- Mentor engineers newer to the profession on ethical dilemmas. Discuss the importance of engineering ethics at your workplace. You can learn a lot by mentoring others and by helping other engineers who have less experience.

Each of the canons in the ASCE Code of Ethics is critically important to the engineering profession and to our individual reputations. Steve Starrett recalls one interesting discussion that was associated with an engineering ethics seminar he conducted for approximately 40 licensed engineers and architects. Steve had described a case where an engineer was being pressured by a supervisor—who was not an engineer—to accept what the engineer knew was an unreasonable risk that a dilapidated bridge was causing to the public. The engineer had just become aware of a two-lane bridge that had extreme deflections during traffic and was about to collapse under its own weight. The engineer immediately closed the bridge to traffic.

The local public thought closure of the bridge for replacement was too much of an inconvenience; the closure caused a 20-minute increase in their commute time. The engineer was confident the risk of collapse was significant and not acceptable by the engineering profession. The supervisor didn't like the public's vocal complaints about the situation and had some inadequately designed temporary support work installed on the bridge. The temporary supports offered minimal improvement to the dilapidated structure. The supervisor overruled the engineer's closure of the bridge by reopening it to traffic once the temporary supports were installed. The public was happy to use the bridge again.

When Steve posed the question, "What would you do if you were the engineer?" to the audience, one individual replied, "I would follow my supervisor's instructions. I have house and car payments to make, and my family depends on me to provide for them."

For the next 20 minutes, one individual after another countered this position by stating, for example, "The engineer is obligated to the public to ensure the bridge is up to safety standards," "The engineer cannot look the other way when the safety of the public is jeopardized," and "The engineer must press on to protect the public."

One engineer in the audience who was near retirement made a particularly passionate statement:

> I have worked as a civil engineer for over 40 years. Considering all of my work, I am most proud of my accomplishments to protect the safety, health, and welfare of the public. I am very glad I advocated a strong position to protect the public when faced with a similar ethical dilemma. Fear over being fired for arguing for the safety of the public is not an acceptable excuse to look the other way. There will always be another job, but you only have one reputation and you have to be able to live with your actions indefinitely.

Drs. Starrett, Bertha, and Lara wish you the very best in your engineering careers. Stand by your moral principles and uphold the engineering codes of ethics to ensure you are fully satisfied upon completion of your engineering careers.

Additional Resources

American Society of Civil Engineers. Code of Ethics website. www.asce.org/code-of-ethics

Engineering ethics. Wikipedia. en.wikipedia.org/wiki/Engineering_ethics

National Society of Professional Engineers. www.nspe.org/resources/ethics

Online Ethics Center for Engineering and Science. National Academy of Engineering. www.onlineethics.org. www.nae.edu/Projects/CEES.aspx

Engineering Ethics. Royal Academy of Engineering. http://www.raeng.org.uk/policy/engineering-ethics/ethics

Engineering Ethics Blog. http://engineeringethicsblog.blogspot.com/

LinkedIn. Ethics-Ethical Professionals. www.linkedin.com/groups/1776046.

Engineering Ethics open access course. Dr. Taft Broome. Massachusetts Institute of Technology IT Open Courseware. http://ocw.mit.edu/courses/engineering-systems-division/esd-932-engineering-ethics-spring-2006/index.htm

National Institute for Engineering Ethics (NIEE). Murdough Center for Engineering Professionalism. Texas Tech University. http://www.depts.ttu.edu/murdoughcenter/

Tara Hoke, ASCE General Counsel. *A Question of Ethics* monthly column. Civil Engineering magazine. American Society of Civil Engineers

Additional Resources and Codes of Ethics

ASCE. Code of Ethics website. www.asce.org/code-of-ethics
ABET. www.abet.org
American Chemical Society. www.acs.org
American Geophysical Union. www.agu.org
American Geosciences Institute. www.americangeosciences.org
American Institute of Architects. www.aia.org
American Institute of Chemical Engineers. www.aiche.org
American Society for Engineering Education. www.asee.org
American Society of Heating, Refrigerating and Air Conditioning Engineers. www.ashrae.org
American Society of Mechanical Engineers. www.asme.org
Association for Computing Machinery. www.acm.org
Engineering Ethics. Royal Academy of Engineering. http://www.raeng.org.uk/policy/engineering-ethics/ethics
Engineering ethics. Wikipedia. en.wikipedia.org/wiki/Engineering_ethics
Engineering Ethics blog. http://engineeringethicsblog.blogspot.com/
Engineering Ethics open access course. Dr. Taft Broome. Massachusetts Institute of Technology IT Open Courseware. http://ocw.mit.edu/courses/engineering-systems-division/esd-932-engineering-ethics-spring-2006/index.htm
IEEE. www.ieee.org
Institute of Industrial and Systems Engineers. www.iise.org
Institute of Transportation Engineers. www.ite.org
Institution of Civil Engineers. www.ice.org.uk
LinkedIn. Ethics-Ethical Professionals. www.linkedin.com/groups/1776046.
National Institute for Engineering Ethics (NIEE). Murdough Center for Engineering Professionalism. Texas Tech University. http://www.depts.ttu.edu/murdoughcenter/

National Society of Professional Engineers. www.nspe.org/resources/ethics

Online Ethics Center for Engineering and Science. National Academy of Engineering. www.onlineethics.org. www.nae.edu/Projects/CEES.aspx

Tara Hoke, JD, ASCE General Counsel. "A Question of Ethics," monthly column. *Civil Engineering* magazine. American Society of Civil Engineers

About the Authors

Dr. Steve Starrett has been on the civil engineering faculty at Kansas State University since 1994. His technical background is water resources engineering. He has long been active in providing engineering ethics education, principally by teaching two graduate-level engineering ethics courses, primarily for distance graduate students. The National Academy of Engineers recognized one of Steve's courses as an exemplar of engineering ethics courses in 2016. Steve has served as the chair of the ASCE Committee on Licensure and Ethics and as a member of the board of the National Institute for Engineering Ethics. Steve is a licensed professional engineer in Kansas and Missouri, and has served as an expert witness in approximately 20 legal disputes. Steve also served as president of the Environmental and Water Resources Institute (EWRI) of ASCE in 2017 and is a Fellow of both ASCE and EWRI.

Steve Starrett touring the Panama Canal.

Dr. Amy Lara is a former associate professor of philosophy at Kansas State University. Amy's research interests are in ethical theory and applied ethics. For six years, she co-taught a research ethics seminar for students in the Physics Research Experience for Undergraduates program at Kansas State. She was also co-principal investigator on a National Science Foundation grant studying the ethics of scientific communication. Currently, she is dividing her time between philosophical research and writing fiction.

Dr. Carlos Bertha is an associate professor of philosophy at the U.S. Air Force Academy in Colorado Springs, CO. He received his bachelor's degree in mechanical engineering from the University of South Florida in 1989. After working at the Savannah District Corps of Engineers for five years, he returned to the University of South Florida, this time to study philosophy. He has been teaching at the Air Force Academy since June 2000. He teaches ethics, analytic philosophy, symbolic logic, and philosophy of science. Carlos' main area of research is engineering ethics, particularly engineering ethics education. He served as a technical advisor for the Global Anti-Corruption, Education and Training Project, an activity supported by ASCE, which produced the movie *Ethicana*. He has been the speaker for a number of National Society of Professional Engineers webinars on engineering ethics. Carlos and Steve as a team have presented short courses in engineering ethics at various national and international events, including ASCE and International Perspective on Water Resource and Environment conferences. Carlos is a colonel in the U.S. Army Reserves, has been deployed to Afghanistan with the Corps of Engineers, and is currently a Defense Support of Civil Authorities (DSCA) officer at the 79th Sustainment Support Command in Los Alamitos, CA.

Dr. Carlos Bertha overseeing a field test during his deployment with the Corps of Engineers in Gardez, Afghanistan

Index